Polymers and Polymer Nanocomposites

Development, Characterization and Applications
(Volume 2)

Related Titles

Biopolymers Based Advanced Materials
ISBN: 978-0-6482205-4-1 (e-book)
ISBN: 978-0-6482205-5-8 (hardcover)

Functional Polymer Blends and Nanocomposites
ISBN: 978-0-6482205-6-5 (e-book)
ISBN: 978-0-6482205-7-2 (hardcover)

Functional Nanomaterials and Nanotechnologies: Applications for Energy & Environment
ISBN: 978-0-6482205-2-7 (e-book)
ISBN: 978-0-6482205-3-4 (softcover)

Advances in Polymer Technology: Material Development, Properties and Performance Evaluation
ISBN: 978-1-925823-00-4 (e-book)
ISBN: 978-1-925823-01-1 (hardcover)

Polymer Nanomaterials for Specialty Applications
ISBN: 978-1-925823-03-5 (e-book)
ISBN: 978-1-925823-04-2 (hardcover)

Advanced Materials
ISBN: 978-1-925823-05-9 (e-book)
ISBN: 978-1-925823-06-6 (hardcover)

Biofuels
ISBN: 978-1-925823-12-7 (e-book)
ISBN: 978-1-925823-13-4 (hardcover)

Liquid Crystalline Polymers
ISBN: 978-1-925823-16-5 (e-book)
ISBN: 978-1-925823-17-2 (hardcover)

Polymer Nanocomposites: Emerging Applications
ISBN: 978-1-925823-14-1 (e-book)
ISBN: 978-1-925823-15-8 (hardcover)

Polymers and Polymer Nanocomposites

Development, Characterization and Applications
(Volume 2)

Dr. Vikas Mittal
Editor and Lead Author

$\mathcal{CWP}$

Central West Publishing

Disclaimer
Every effort has been made by the publisher, editor and authors while preparing this book, however, no warranties are made regarding the accuracy and completeness of the content. The publisher, editor and authors disclaim without any limitation all warranties as well as any implied warranties about sales, along with fitness of the content for a particular purpose. Citation of any website and other information sources does not mean any endorsement from the publisher and authors. For ascertaining the suitability of the contents contained herein for a particular lab or commercial use, consultation with the subject expert is needed. In addition, while using the information and methods contained herein, the practitioners and researchers need to be mindful for their own safety, along with the safety of others, including the professional parties and premises for whom they have professional responsibility. To the fullest extent of law, the publisher, editor and authors are not liable in all circumstances (special, incidental, and consequential) for any injury and/or damage to persons and property, along with any potential loss of profit and other commercial damages due to the use of any methods, products, guidelines, procedures contained in the material herein.

NATIONAL LIBRARY OF AUSTRALIA

A catalogue record for this book is available from the National Library of Australia

ISBN (print): 978-1-925823-71-4

Contents

Preface

A wide spectrum of polymer based materials with optimal property profiles and performance have resulted owing to the recent advancements in polymer science and technology. The current book presents various aspects related to the development and characterization of a range of such polymeric systems and polymer nanocomposites with special emphasis on their applications.

Chapter 1 presents the thermo-mechanical properties of high density polyethylene/polyamide blend nanocomposites with varying amounts of carbon nanotubes, so as to further develop this important class of blend composites for various applications. The mechanical, thermal and structural characteristics of graphene nanoplatelets filled polyethylene/polyamide blend nanocomposites have also been analyzed in Chapter 2. In Chapter 3, novel aerogels have been prepared using nanofibrillated cellulose by employing vacuum drying. Graphene-based carbonaceous monolithic material containing cobalt ferrite nanoparticles has been synthesized for adsorption applications in Chapter 4. To explore the usefulness of metal organic frameworks (HKUST-1) for generating high performance membranes, mixed matrix membranes involving PEBAX and HKUST-1 (in varying weight fractions) were developed and analyzed for their mechanical and structural characteristics in Chapter 5. In Chapter 6, the preparation and characterization of bio-nanocomposites, obtained through solution blending of bio-polyamide and doped graphene (n and p), have been presented. Chapter 7 reviews different anti-microbial polymeric materials and coatings, whereas different living polymerization techniques have been briefly presented in Chapter 8. In Chapter 9, the development of polymer nanocomposite inks and pigments has been briefly assessed.

The book would not have been successfully accomplished without the support of chapter contributors. The book is dedicated to my family for unswerving support, constant motivation and constructive suggestions for improvement.

Vikas MITTAL

1

CNTs Reinforced Polyethylene/Polyamide Blend Nanocomposites: Structural Characterization and Properties

1.1 Introduction

Polymer blends have received significant research interest in the past few years because of the ideal balance of cost and properties. For obtaining the blends possessing improved properties, the advantageous selection of the blend components and the control of the microstructure developed at the time of blending are required [1]. In addition to the blending of distinct polymers, an alternate method for enhancing the polymer performance is the incorporation of solid particles by melt mixing the components. Utilization of the nanosized particles permits an exceptional reduction in the quantity of filler necessary to enhance the properties as compared to the typical micron-sized particles [2]. Among the different immiscible polymer blends of technological significance, several research studies have concentrated on the blends of polyamides (PA) and polyethylene (PE), the polymers broadly utilized in a wide variety of automotive and packaging products due to their effective mechanical properties, barrier performance and processability. Polyamides act as a superior barrier to oxygen and offer excellent mechanical properties, whereas polyethylene performs as a good barrier to humidity and contributes better impact resistance.

Recently, the biopolymers have received research attention due to the increasing consciousness towards the usage of environmentally beneficial materials [3]. The biopolymers have replaced the conventional polymers in various industrial, automotive, packaging, electronic/electrical and consumer good applications. Presently, the greatest necessity for the bio-based polymers is in the packaging sector, where there are remarkable opportunities for the biopolymers as engineering resins. Nevertheless, the biopolymers are most-

Haleema Saleem and Vikas Mittal*, The Petroleum Institute (part of Khalifa University of Science and Technology), Abu Dhabi, UAE
*Current address: Bletchington, Wellington County, Australia

ly expensive and possess poor property profiles, as compared to several commercial thermoplastic polymers. Therefore, the biopolymers need improvement in their properties for effectively competing with the conventional polymers. Polyamides are a category of synthetic engineering polymers, comprising of the amide group repeating units in the polymer chains [4,5]. Most of the commercial PAs are based on the petroleum-based monomers. Nonetheless, during past several decades, a developing interest in the development and commercialization of PAs based on bio-based monomers has been noticed. The most important bio-based commercial PAs are PA1010, PA11, PA610 and PA1012. PA1010 or Nylon 1010 is a commercially developed engineering PA [6] possessing several unique characteristics like high toughness, abrasion resistance, strength and elasticity [7]. Nevertheless, it has a few disadvantages such as moisture adsorption, which can generally be overcome by developing the blends as well as nanocomposites of PA1010 [8].

As mentioned earlier, there has been a developing interest in the blends of polyolefins and PAs in the recent years. When appropriately blended, the polyolefin and PA blends can conceivably contribute an extensive variety of desirable qualities, for example, low cost, lesser water sorption and greater chemical resistance. The mechanical characteristics of the aforementioned blends are strongly associated with the ratio of PA to polyolefin [9]. The blending of polyolefin with PA also enhances the impact resistance. As also mentioned earlier, the inclusion of a comprehensive range of fillers in the conventional polymers has been noted to result in significant enhancements in mechanical, gas barrier, thermal and rheological characteristics [10]. Nanofillers such as graphene, carbon nanotubes (CNTs) and nanoclays can develop the filler networks inside the polymer matrices at lower loadings due to the greater aspect ratios [11-13]. In order to generate high value materials from the biopolymers, corresponding nanocomposites employing the fillers such as layered silicates, nanotubes, etc., have been reported [14]. Polyamide nanocomposites have also obtained significant attention due to exceptional qualities such as effective barrier performance, modulus, electrical conductivity as well as non-flammability [15,16].

The CNTs, a one-dimensional (1-D) carbonaceous material, have brought about remarkable activities in various territories of engineering and science due to their exclusive physico-chemical properties. The diameter in CNTs is at the nanometer scale, and these can be either multi-walled (MWCNT) or single-walled (SWCNT). Due to their high aspect ratio, stiffness and strength at lower thickness, the CNTs have drawn the attention as reinforcing fillers for the development of low-weight and superior performance polymer based materials [17]. The methods for the dispersion and distribution of the CNTs inside the polymer materials play a crucial role in achieving the properties of the final product. For the preparation of CNTs filled polymer nanocomposites, different methods such as solution casting, melt mixing and *in-situ* polymerization have been utilized.

As reported earlier, the polyolefin/PA blend systems display excellent impact strength [18-20], along with high mechanical, thermal and oil resistance properties [21,22]. The polyolefin/PA blends have remarkable importance for the use in food packaging because of the superior moisture barrier characteristics of the polyolefin component and the excellent oxygen barrier properties of the PA phase [23]. The influence of CNTs on the morphology, electrical and mechanical properties of the immiscible polyethylene/polypropylene (PE/PP) blends was also examined by Al-Saleh [24]. The authors noted significant increase in the tensile strength with an increment in the CNT concentration. Also, the electrical conductivity and mechanical properties of the CNTs reinforced polyamide 6 and its blends with acrylonitrile/butadiene/styrene(ABS) were reported by Meincke *et al.* [25]. It was noted that the CNTs-filled polymer blends displayed excellent properties in Izod notched impact and tensile tests. The nanocomposites exhibited an enhancement of 27% in the Young's modulus, nevertheless, the elongation at break of the aforementioned nanocomposites significantly reduced due to the embrittlement of polyamide 6.

In this study, the thermo-mechanical properties of high density polyethylene (HDPE)/PA1010 blend nanocomposites with CNTs have been studied, at varying amounts of CNTs (1, 2, 3 and 5 wt%), so as to further develop this important class of blend composites for various applications.

1.2 Experimental

1.2.1 Materials

Renewably sourced biodegradable polyamide PA1010, having the trade name Zytel® RS LC1200 BK385 (70 wt% renewably sourced ingredient), was provided by DuPont. It had a melt viscosity of 500 Pa.s at 220 °C and density of 1.03 g/cm^3. This material was UV and heat stabilized and was suitable for multiple extrusion applications. HDPE grade PE BB2581 was provided by Abu Dhabi Polymers Company Limited. The MWCNTs utilized in the study were commercially procured and had outside diameter <8 nm, inside diameter 2-5 nm, purity >95%, length 10-30 um and SSA >500 m²/g.

1.2.2 Preparation of HDPE/PA1010 Blend

HDPE/PA1010 blend (1: 1 ratio) was generated by the melt mixing technique. HDPE (2 g) and PA1010 (2 g) were added to a twin conical screw extruder (Thermo Scientific Haake Minilab II). A mixing temperature of 220 °C was utilized, and the mixing was carried out for 5 min at 100 rpm in a co-rotating mode, with a batch size of 4 g.

1.2.3 Preparation of HDPE/PA1010/CNTs Nanocomposites

HDPE/PA1010/CNT nanocomposites were prepared by the melt mixing technique, using suitable amount of HDPE, PA1010 and MWCNTs nanofiller employing a twin conical screw extruder (Thermo Scientific Haake Minilab II). A mixing temperature of 220 °C was utilized, and the mixing was carried out for 5 min at 100 rpm in a co-rotating mode, with a batch size of 4 g. A series of HDPE/PA-1010/CNTs nanocomposites were prepared by varying the content of MWCNTs as 1, 2, 3 and 5 wt%, as presented in Table 1.1.

Table 1.1 Composition of CNTs based HDPE/PA1010 nanocomposites

Blend/Nanocomposites	HDPE (%)	PA1010 (%)	CNTs (%)
HDPE+ PA1010 blend	50.0	50.0	0.0
HDPE/PA1010+1% CNTs	49.5	49.5	1.0
HDPE/PA1010+2% CNTs	49.0	49.0	2.0
HDPE/PA1010+3% CNTs	48.5	48.5	3.0
HDPE/PA1010+5% CNTs	47.5	47.5	5.0

After the extrusion process, the injection molding of the specimens was carried out in an injection molding machine (Thermo Scientific Haake Minijet Pro). The cylinder and mold temperatures were fixed as 235 °C and 125 °C, respectively. Also, the post and injection pressures were set as 470 and 570 bars, respectively, for a time period of 10 s. Various sample shapes, such as dumbbells and circular discs, were generated for each composition. The blend and nanocomposites were designated as PEPA (HDPE/PA1010 blend),

PEPACNT1 (HDPE/PA1010+1 wt% CNTs), PEPACNT2 (HDPE/PA1010+2 wt% CNTs), PEPACNT3 (HDPE/PA1010+3 wt% CNTs) and PEPACNT5 (HDPE/PA1010+5 wt% CNTs).

1.2.4 Characterization

Crystallization behavior of PEPA blend and PEPA blend nanocomposites was analyzed using differential scanning calorimetry (DSC) under N_2 atmosphere. The DSC thermograms were obtained utilizing a TA Discovery DSC with the help of TRIOS software. The extruded specimens of almost 3-7 mg weight were employed for the characterization. The tests were performed under non-isothermal conditions. Two heating and cooling cycles were carried out. During the first cycle, the specimens were heated from 25 °C to 275 °C, and subsequently cooled to -30 °C at a rate of +/- 10 °C/min for analyzing the re-crystallization temperature. Later, in the second cycle, the specimens were again heated to 275 °C for recording the melt temperature, and subsequently cooled to -30 °C at a rate of +/- 10 °C/min. The peak melting temperature (T_m), enthalpy of melting (ΔH_m), peak crystallization temperature (T_C) and enthalpy of crystallization (ΔH_C) were analyzed.

The crystalline structure of the PEPA blend and nanocomposites was analyzed using wide angle X-ray diffraction (WAXRD). Panalytical X'Pert Pro diffractometer was employed for obtaining the WAXRD spectra of the injection molded specimens, with the step size of 0.020° s^{-1} and a scanning angle range of 2θ = 5-40°. The power setting of the XRD apparatus were 40 mA and 40 kV in reflection mode.

Attenuated total reflection-Fourier transform infrared spectroscopy (ATR-FTIR) analysis of the pristine CNTs, pure PA1010, HDPE/PA1010 blend and blend nanocomposites was carried out using the injection molded samples by employing a Bruker VERTEX 70 spectrometer, equipped with a DLaTGS detector with KBr window, diamond ATR accessory and KBr beam-splitter. The ATR accessory had a ZnSe crystal with a refractive index of about 2.4@1000 cm^{-1} and a depth penetration of almost 2 μm at 45° angle of incidence. The IR spectra in the 4000-600 cm^{-1} range were plotted by employing the OPUS software at 8 cm^{-1} resolution with an average scans of 64.

The elongation at break, tensile stress at yield and tensile modulus of the PEPA blend and nanocomposites were analyzed using Instron 5567 universal testing machine at the strain rate of 10 mm/min. The analysis was carried out at room temperature in accordance with ASTM D 638. The dumbbell-shaped samples of 4 mm width, 2 mm thickness and 53 mm length were employed. For the evaluation of the mechanical properties, Win Test Analysis software was employed. An average of five values was reported.

The Raman spectroscopic analysis was carried out utilizing the confocal microRaman spectrometer Horiba LabRAM HR, equipped with a 514 nm laser source.

The thermal conductivity analysis of the blend and nanocomposites was carried out utilizing the laser flash equipment LFA 447 from Netzsch, Germany (operated with the assistance of Nanoflash software). This technique is appropriate for conducting the non-contact measurements of thermal diffusivity, heat capacity and thermal conductivity of different materials. Thermal conductivity (λ) of the samples was obtained using the relation $\lambda = \alpha.\rho.Cp$ where α, ρ and Cp are the thermal diffusivity, density and specific heat of the samples, respectively.

1.3 Results and Discussion

Calorimetric properties of pristine PA1010, PEPA blend and blend nanocomposites are outlined in Table 1.2. Figures 1.1 and 1.2 also illustrate the melting and cooling curves of these samples. The pristine PA1010 showed two melting peaks, one at 191 °C and the other at 199 °C. The observed multiple melting behaviors might be due to the re-crystallization during the DSC analysis. PA1010 is comprised of crystallite dispersion of different levels of perfection, which gets

Table 1.2 Calorimetric behavior of PEPA blend and PEPA blend nanocomposites

Composition	Heating			Cooling		
	ΔH_m (cal/g)	ΔH_m (J/g)	T_m (°C)	ΔH_c (cal/g)	ΔH_c (J/g)	T_c (°C)
Pure PA1010	6.3	26.4	191,199	10.6	44.2	177
PEPA blend	1.4	5.8	199	4.2	17.5	178
PEPACNT1	5.3	22.3	198	4.6	19.4	183
PEPACNT2	4.8	20.0	198	4.8	20.0	183
PEPACNT3	5.2	21.6	198	5.2	21.6	183
PEPACNT5	1.2	5.0	198	3.7	15.5	182

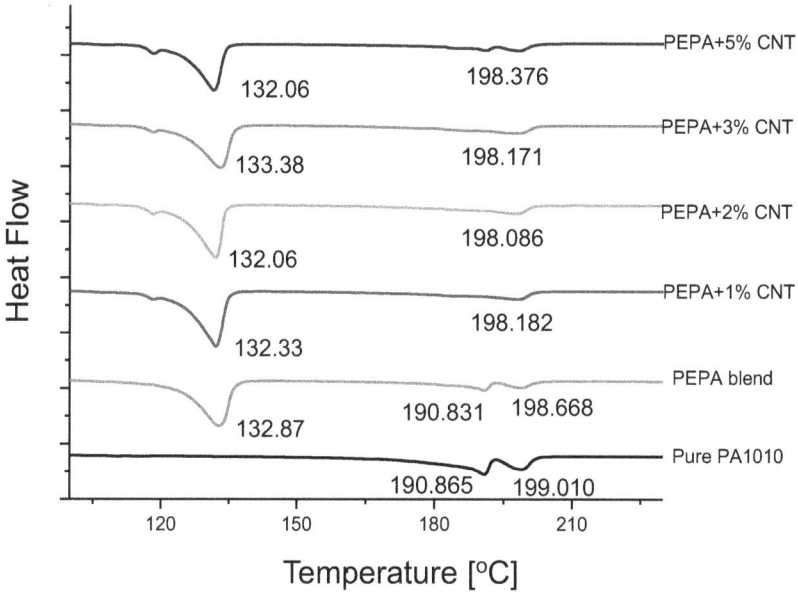

Figure 1.1 DSC melting curves of PEPA blend and nanocomposites.

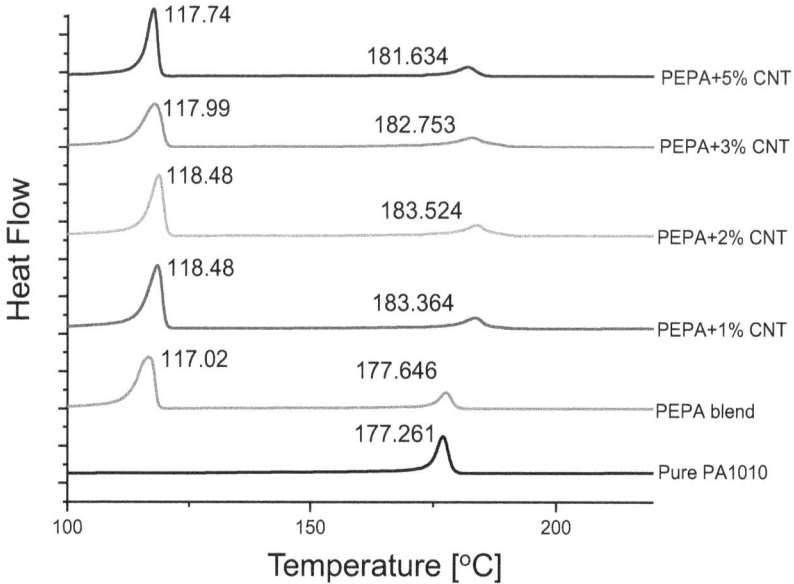

Figure 1.2 DSC cooling curves of PEPA blend and nanocomposites.

actively influenced by its previous thermal history. The peak located at 199 °C is the melting of highly stable crystals, whereas the peak at 191 °C might be attributed to the melting of unstable crystals [26]. The melting temperature and enthalpy obtained for HDPE were 133 °C and 21.7 cal/g, respectively. With the inclusion of the CNTs, the melting temperature of the PEPA blend composites exhibited an unchanged value of 198 °C for the PA phase and 132 °C for the PE phase. Conversely, an increase of ΔH_m (PA phase) could be observed for the CNTs filled PEPA nanocomposites till 3 wt%, relative to PEPA. This confirmed that the CNTs in the blend promoted the polymer crystallization. The crystallization temperatures of PEPA blend (both PA and PE phases) also increased slightly with the incorporation of CNTs, with the maximum value for 2% CNTs content.

Figure 1.3 presents the WAXRD patterns of pristine PA1010, pure HDPE, pure CNTs, PEPA blend and nanocomposites. HDPE exhibited two intense diffraction peaks, a strong reflection at 2θ = 24.3° and a weaker peak at 21.8°, corresponding to the (200) and (110) crystal planes, respectively. A strong Bragg peak concentrated around 0.34 nm, associated with the inter-shell spacing within the

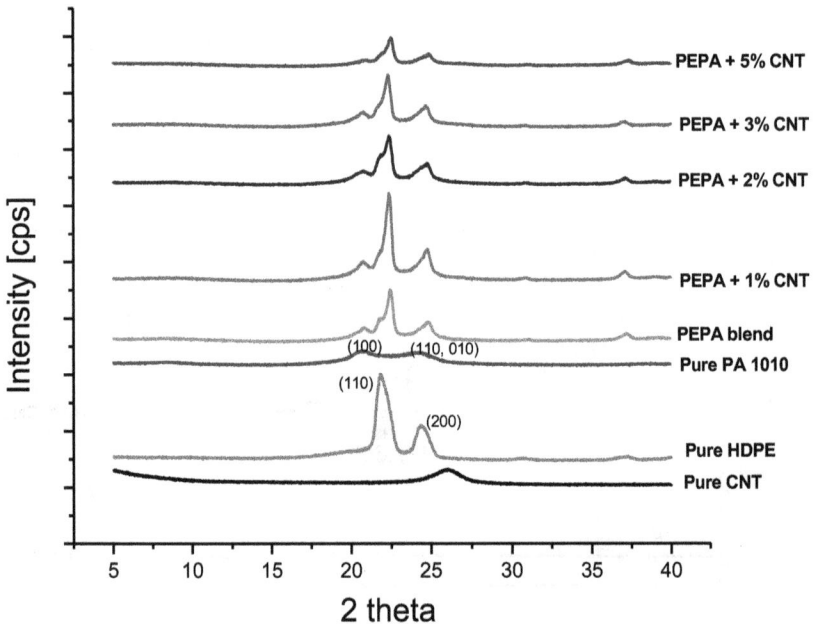

Figure 1.3 XRD patterns of pure PA1010, pure HDPE, pure CNTs, PEPA blend and PEPA blend nanocomposites.

nanotubes, was observed in the diffraction pattern of the MWCNTs. The WAXRD spectrum of the pristine PA1010 exhibited two distinct diffraction signals, at 2θ = 24.4° and 20.5°, associated with the (110/010) and 100 planes, respectively, of the α-phase crystal [27,28]. In the case of PEPA blend, the diffraction peak of (110) crystal plane of HDPE overlapped with the PA1010 diffraction peak of (100) at 2θ = 22.3°, and the diffraction peak of (200) crystalline plane of HDPE overlapped with the PA1010 diffraction peak of (110/010) at 2θ = 24.7°. XRD spectra nanocomposites exhibited a slight enhancement in the peak positions, though there was no specific effect of CNT concentration.

As the infrared (IR) bands in polymers are sensitive to molecular conformations, chain packing framework and hydrogen bonding, the IR analysis was carried out for examining the structural changes in the HDPE/PA1010 blend by the incorporation of CNTs. Pristine PA1010 exhibited the N-H stretching modes (amide A) of hydrogen bonded N-H groups at 3305 cm^{-1} and the carbonyl stretching vibrations (amide I) of hydrogen bonded carbonyl groups at 1636 cm^{-1} [29,30]. The coupling of C-N stretching and N-H in-plane bending modes (amide II) was observed at 1538 cm^{-1}. From Figure 1.4, it was

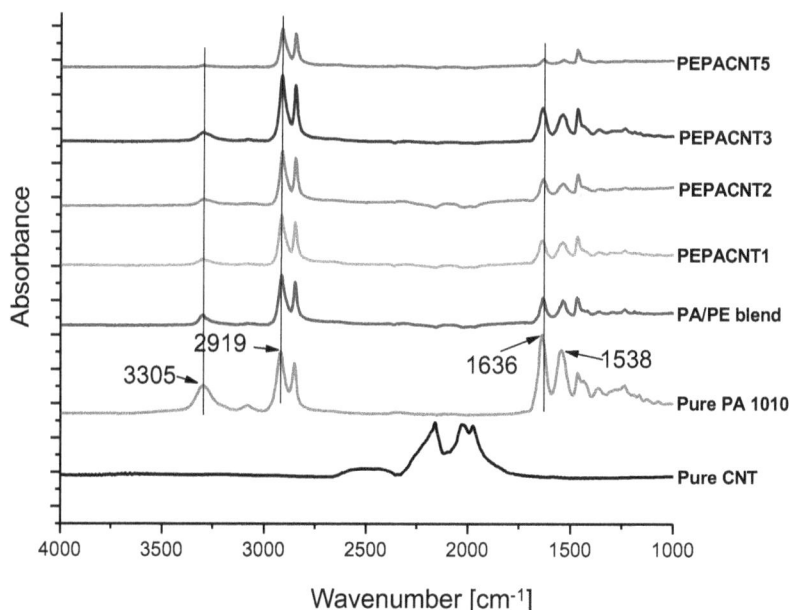

Figure 1.4 FTIR spectra of pure PA1010, pure CNTs, PEPA blend and PEPA blend nanocomposites.

noted that the carbonyl stretching vibrations (amide I) of hydrogen bonded carbonyl groups at 1636 cm^{-1} in the nanocomposites showed a decrease in intensity, relative to the HDPE/PA1010 blend. It was also quantitatively analyzed by determining the area ratio of this band (A_{1636}) to the C-H stretching vibration band seen at 2919 cm^{-1}. As presented in Table 1.3, the A_{1636}/A_{2919} ratio for HDPE/PA1010 blend was higher than the HDPE/PA1010 blend nanocomposites, clearly revealing that the particular amide bonds in the disordered amorphous area of the HDPE/PA1010 blend were subjected to chemical interactions in the presence of CNTs.

Table 1.3 A_{1636}/A_{2925} for pure PA1010, PEPA and nanocomposites

Material	A_{1636}	A_{2919}	A_{1636}/A_{2925}
Pure PA1010	0.476	0.368	1.29
PEPA blend	0.202	0.331	0.60
PEPACNT1	0.188	0.324	0.58
PEPACNT2	0.225	0.392	0.57
PEPACNT3	0.209	0.394	0.53
PEPACNT5	0.065	0.241	0.27

The mechanical properties of the blend nanocomposites are associated with different factors inclusive of the blend properties, impact of the nanofiller on the host polymer, blend morphology, positioning of the nanofiller inside the blend and influence of the filler on the state of adhesion at the blend's interface. The tensile properties of the CNTs filled HDPE/PA1010 nanocomposites are presented in Table 1.4. It was noted that the CNTs significantly enhanced the Young's modulus of the HDPE/PA1010 blend (Figure 1.5). For instance, the tensile modulus was increased to 1743 MPa for 5% CNTs filled HDPE/PA1010 nanocomposite, thus, showing an increment of 39% as compared to the HDPE/PA1010 blend. The molecular architecture as well as phase behavior perform a critical role in defining the chain orientation and tensile strength. The unfilled PA1010 had a tensile strength of about 50 MPa. The HDPE/PA1010 blend exhibited a lower tensile strength relative to pristine PA1010. The addition of CNTs gradually decreased the tensile strength to 26 MPa for 1% CNTs incorporation and 20 MPa in the case of 5% CNTs incorporation. The reduction in the strength can be attributed to the incompatible blend morphology. In addition, a gradual decrease in the elongation at break was noticed on incorporating the filler, thus, indicating enhanced brittleness.

Table 1.4 Tensile properties of PA1010, HDPE/PA1010 blend and nanocomposites

Material	Elongation at break (%)	Tensile stress at yield (MPa)	Tensile modulus (MPa)
Pure PA1010	19.8	50	1084
PEPA blend	7.2	28	1254
PEPACNT1	3.9	26	1425
PEPACNT2	2.9	23	1460
PEPACNT3	2.7	22	1479
PEPACNT5	1.9	20	1743

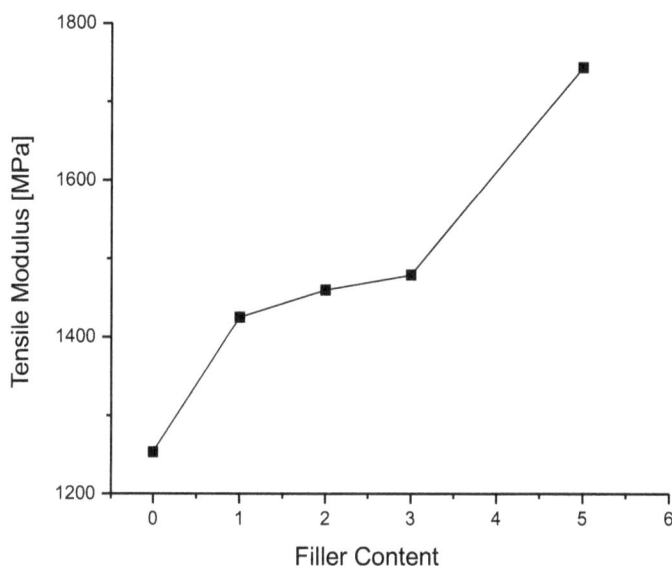

Figure 1.5 Tensile modulus of HDPE/PA1010 blend and different CNT based HDPE/PA1010 nanocomposites.

In Figure 1.6, the Raman spectrum of pure CNTs exhibited two typical graphite bands at 1574 cm^{-1} (G band due to in-plane vibrations of sp^2 bonded carbon atoms) and 1320 cm^{-1} (D band related to sp^3 bonds). An additional band, known as the G" band, was also present in the Raman range at 2642 cm^{-1} due to the D band overtone [31]. The spectrum of HDPE (Figure 1.7) exhibited conventional Raman modes of PE. The Raman bands observed at ~1069 cm^{-1} and ~1137 cm^{-1} could be attributed to the C-C stretching modes in the

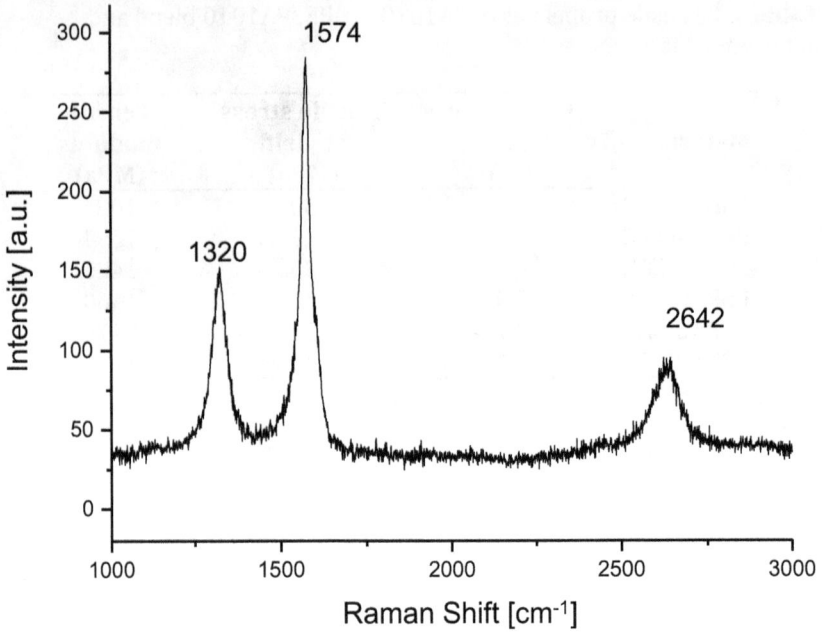

Figure 1.6 Raman spectrum of pristine CNTs.

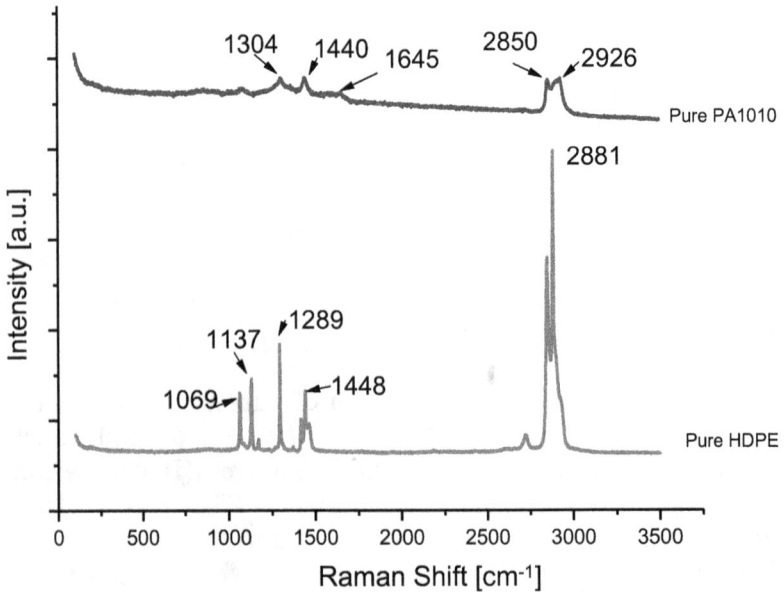

Figure 1.7 Raman spectra of pure HDPE and pure PA1010.

crystalline and anisotropic phases of PE. PA1010 spectrum consisted of the characteristic bands of PA, where the bands seen at 1645 cm^{-1} (amide I) and 1304 cm^{-1} (amide III) were associated with the presence of the amide groups in the polymer. The major bands observed at 2850 cm^{-1} and 2926 cm^{-1} were associated with the symmetric and asymmetric CH$_2$ stretching vibrations, respectively. Also, from Figure 1.8, it was observed that the Raman spectrum of 5 wt% CNTs filled HDPE/PA1010 nanocomposite was moderately influenced by the CNT bands. The analysis of Raman I_D/I_G ratio (I_D – Raman intensity of D peak; I_G – Raman intensity of G peak) of pure CNTs and of 5 wt% CNTs filled HDPE/PA1010 nanocomposite was performed to determine the interactions between the nanofiller and blend, as presented in Table 1.5. For the nanocomposite with 5% CNTs content, the value of I_D/I_G was greater, relative to MWCNTs. The observed higher I_D/I_G ratio of the PEPACNT5 nanocomposite indicated the possible generation of covalent interaction between the HDPE/PA1010 blend and MWCNTs, which moderately breaks the sp^2 carbon framework and subsequently generates a stronger interaction between them.

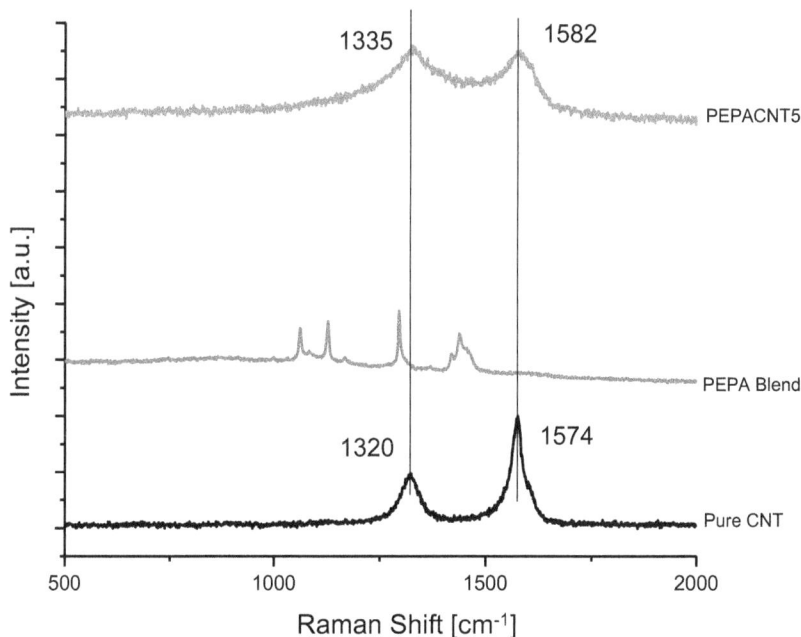

Figure 1.8 Raman spectra of pure CNTs, PEPA blend and 5 wt% CNTs filled HDPE/PA1010 nanocomposite.

Table 1.5 Raman intensity ratio (I_D/I_G) of pristine CNT and 5 wt% CNTs filled nanocomposite

Material	D band (cm^{-1})	G band (cm^{-1})	I_D	I_G	Raman Intensity ratio (I_D/I_G)
Pure CNTs	1320	1574	0.480	0.993	0.483
PEPACNT5	1335	1582	0.986	0.951	1.036

The thermal conductivity of suspended graphene is approximately 5000 W/m.K [32], whereas pristine HDPE and PA1010 have low conductivity. Thus, the outcome of the incorporation of CNTs on the thermal conductivity of HDPE/PA1010 blend was analyzed, as presented in Figure 1.9 and Table 1.6. The presence of CNTs enhanced the thermal conductivity of the HDPE/PA1010 blend significantly, as a function of filler content. Also, the conductivity was observed o decrease with temperature. At a temperature 60 °C, the composites with 3 and 5 wt% filler content attained the values 0.21 W/m.K and 0.236 W/m.K respectively, indicating 30% and 47% enhancement relative to the HDPE/PA1010 blend.

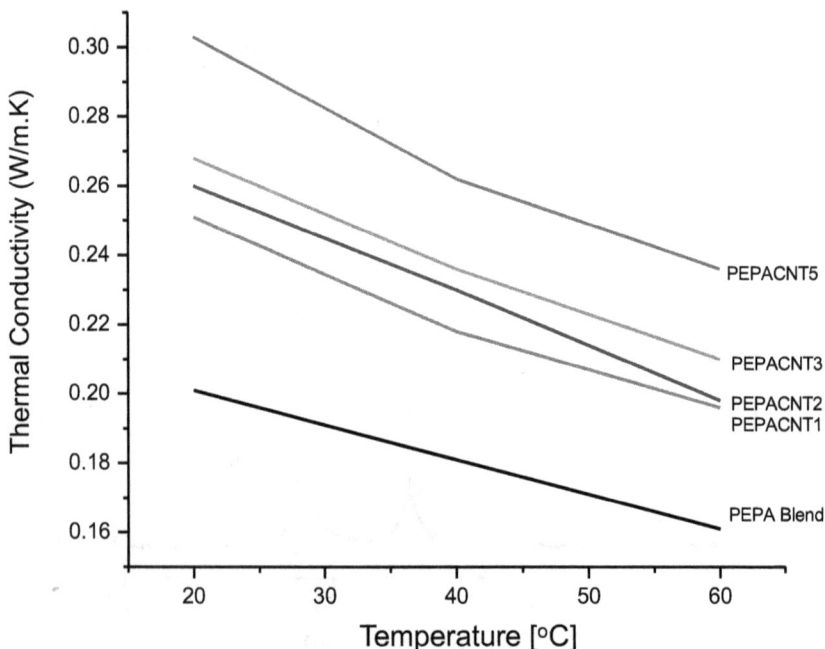

Figure 1.9 Thermal conductivity analysis of PEPA blend and PEPA blend nanocomposites.

Table 1.6 Thermal conductivity of PEPA blend and PEPA blend nanocomposites

Temp.	PEPA Blend	PEPACNT1	PEPACNT2	PEPACNT3	PEPACNT5
20 °C	0.201	0.251	0.260	0.268	0.303
40 °C	0.181	0.218	0.230	0.236	0.262
60 °C	0.161	0.196	0.198	0.210	0.236

1.4 Conclusions

In this study, the thermo-mechanical properties and structural characteristics of polyethylene (HDPE)/polyamide1010 (PA1010) nanocomposites containing the multi-walled carbon nanotubes (MWCNTs) have been analyzed. The inclusion of the CNTs did not affect the peak melting temperatures, though a slight increase in ΔH_m could be observed. The diffraction peaks of the blend also exhibited a slight increase in the 2θ values in the nanocomposites. The A_{1636}/A_{2919} ratio from the FTIR spectra was greater for HDPE/PA1010 blend, relative to the nanocomposites, indicating that the particular amide bonds in the disordered amorphous area of HDPE/PA1010 blend were subjected to chemical interaction in the presence of CNTs. CNTs significantly enhanced the Young's modulus of the HDPE/PA1010 blend, with the composite with 5% CNTs showing an increment of 39%. CNTs also enhanced the thermal conductivity of the HDPE/PA1010 blend significantly, as a function of filler content.

References

1. *Micro-and Nanostructured Multiphase Polymer Blend Systems: Phase Morphology and Interfaces*, Harrats, C., Thomas, S., and Groeninckx, G. (eds.), CRC Press, USA (2005).
2. Jordan, J., Jacob, K. I., Tannenbaum, R., Sharaf, M. A., and Jasiuk, I. (2005) Experimental trends in polymer nanocomposites - A review. *Materials Science and Engineering A*, **393**(1), 1-11.
3. Bordes, P., Pollet, E., and Avérous, L. (2009) Nano-biocomposites: biodegradable polyester/nanoclay systems. *Progress in Polymer Science*, **34**(2), 125-155.
4. Marchildon, K. (2011) Polyamides–still strong after seventy years. *Macromolecular Reaction Engineering*, **5**(1), 22-54.

5. Carraher, Jr., C. E., Morrison, A., Roner, M. R., Moric-Johnson, A., Al-Huniti, M. H., and Miller, L. (2015) Metallocene-containing polyesters from reaction of 3, 5-pyridinedicarboxylic acid and metallocene dihalides and their preliminary ability to inhibit cancer cell growth. *Journal of the Chinese Advanced Materials Society*, **3**(4), 310-327.

6. Zhishen, M., Qingbo, M., Jinhua, F., Hongfang, Z., and Donglin, C. (1993) Crystal structure and thermodynamic parameters of nylon-1010. *Polymer International*, **32**(1), 53-60.

7. Zhang, S. L., Wang, G. B., Jiang, Z. H., Wang, D., Ma, R. T., and Wu, Z. W. (2005) Impact properties, phase structure, compatibility, and fracture morphology of polyamide-1010/thermoplastic poly (ester urethane) elastomer blends. *Journal of Polymer Science, Part B: Polymer Physics*, **43**(10), 1177-1185.

8. Lu, Z., Wu, A., Ou, X., Zhang, S., Niu, J., Ji, S., and Ling, Y. (2017) Enhanced anti-aging and mechanical properties of polyamide 1010 by sol-hydrothermal synthetic titanium dioxide-coated kaolinite addition. *Journal of Alloys and Compounds*, **693**, 381-388.

9. Gonzalez-Montiel, A., Keskkula, H., and Paul, D. R. (1995) Impact-modified nylon 6/polypropylene blends: 1. Morphology-property relationships. *Polymer*, **36**(24), 4587-4603.

10. Mittal, V. (2008) Mechanical and gas permeation properties of compatibilized polypropylene–layered silicate nanocomposites. *Journal of Applied Polymer Science*, **107**(2), 1350-1361.

11. Carretero-González, J., Retsos, H., Verdejo, R., Toki, S., Hsiao, B. S., Giannelis, E. P., and López-Manchado, M. A. (2008) Effect of nanoclay on natural rubber microstructure. *Macromolecules*, **41**(18), 6763-6772.

12. Wang, K., Liang, S., Deng, J., Yang, H., Zhang, Q., Fu, Q., Dong, X., Wang, D., and Han, C. C. (2006) The role of clay network on macromolecular chain mobility and relaxation in isotactic polypropylene/organo-clay nanocomposites. *Polymer*, **47**(20), 7131-7144.

13. Bréchet, Y., Cavaille, J. Y., Chabert, E., Chazeau, L., Dendievel, R., Flandin, L., and Gauthier, C. (2001) Polymer based nanocomposites: Effect of filler-filler and filler-matrix interactions. *Advanced Engineering Materials*, **3**(8), 571-577.

14. Favier, V., Cavaille, J. Y., Canova, G. R., and Shrivastava, S. C. (1997) Mechanical percolation in cellulose whisker nanocomposites. *Polymer Engineering and Science*, **37**(10), 1732-1739.

15. Ma, L., Niu, H., Cai, J., Zhao, P., Wang, C., Lian, Y., Bai, X., and Wang, W. (2014) Optical, electrochemical, photoelectrochemical and electrochromic properties of polyamide/graphene oxide with various feed ratios of polyamide to graphite oxide. *Journal of Materials Chemistry C*, **2**(12), 2272-2282.

16. Zheng, D., Tang, G., Zhang, H. B., Yu, Z. Z., Yavari, F., Koratkar, N.,

Lim, S. H., and Lee, M. W. (2012) In situ thermal reduction of graphene oxide for high electrical conductivity and low percolation threshold in polyamide 6 nanocomposites. *Composites Science and Technology*, **72**(2), 284-289.

17. Ganesan, Y., Peng, C., Lu, Y., Loya, P. E., Moloney, P., Barrera, E., Yakobson, B. I., Tour, J. M., Ballarini, R., and Lou, J. (2011) Interface toughness of carbon nanotube reinforced epoxy composites. *ACS Applied Materials and Interfaces*, **3**(2), 129-134.

18. Utracki, L. A., Dumoulin, M. M., and Toma, P. (1986) Melt rheology of high density polyethylene/polyamide-6 blends. *Polymer Engineering and Science*, **26**(1), 34-44.

19. Chen, C. C., Fontan, E., Min, K., and White, J. L. (1988) An investigation of instability of phase morphology of blends of nylons with polyethylenes and polystyrenes and effects of compatibilizing agents. *Polymer Engineering and Science*, **28**(2), 69-80.

20. Hobbs, S. Y., Bopp, R. C., and Watkins, V. H. (1983) Toughened nylon resins. *Polymer Engineering and Science*, **23**(7), 380-389.

21. Willis, J. M., and Favis, B. D. (1988) Processing-morphology relationships of compatibilized polyolefin/polyamide blends. Part I: The effect of an Ionomer compatibilizer on blend morphology. *Polymer Engineering and Science*, **28**(21), 1416-1426.

22. Koulouri, E. G., Georgaki, A. X., and Kallitsis, J. K. (1997) Reactive compatibilization of aliphatic polyamides with functionalized polyethylenes. *Polymer*, **38**(16), 4185-4192.

23. Holsti-Miettinen, R. M., Perttilä, K. P., Seppälä, J. V., and Heino, M. T. (1995) Oxygen barrier properties of polypropylene/polyamide 6 blends. *Journal of Applied Polymer Science*, **58**(9), 1551-1560.

24. Al-Saleh, M. H. (2016) Carbon nanotube-filled polypropylene/polyethylene blends: compatibilization and electrical properties. *Polymer Bulletin*, **73**(4), 975-987.

25. Meincke, O., Kaempfer, D., Weickmann, H., Friedrich, C., Vathauer, M., and Warth, H. (2004) Mechanical properties and electrical conductivity of carbon-nanotube filled polyamide-6 and its blends with acrylonitrile/butadiene/styrene. *Polymer*, **45**(3), 739-748.

26. Mittal, V., Chaudhry, A. U., and Luckachan, G. E. (2014) Biopolymer–thermally reduced graphene nanocomposites: structural characterization and properties. *Materials Chemistry and Physics*, **147**(1), 319-332.

27. Xu, Z., and Gao, C. (2010) In situ polymerization approach to graphene-reinforced nylon-6 composites. *Macromolecules*, **43**(16), 6716-6723.

28. Li, Y., and Yan, D. (2001) Preparation, characterization and crystalline transition behaviors of polyamide 4 14. *Polymer*, **42**(11), 5055-5058.

29. Yoshioka, Y., Tashiro, K., and Ramesh, C. (2003) Structural change

in the Brill transition of Nylon m/n (2) conformational disordering as viewed from the temperature-dependent infrared spectral measurements. *Polymer*, **44**(20), 6407-6417.

30. Yoshioka, Y., and Tashiro, K. (2003) Structural change in the Brill transition of Nylon m/n (1) Nylon 10/10 and its model compounds. *Polymer*, **44**(22), 7007-7019.

31. Bokobza, L., and Zhang, J. (2012) Raman spectroscopic characterization of multiwall carbon nanotubes and of composites. *Express Polymer Letters*, **6**(7), 601-608.

32. Lehman, J. H., Terrones, M., Mansfield, E., Hurst, K. E., and Meunier, V. (2011) Evaluating the characteristics of multiwall carbon nanotubes. *Carbon*, **49**(8), 2581-2602.

2

Graphene Reinforced Polyamide/Polyethylene Blend Nanocomposites: Structural Characterization and Properties

2.1 Introduction

The morphology of the blends relies upon various parameters like viscosity ratio, composition and interfacial tension of the component phases as well as processing conditions at the time of blending [1-4]. Apart from the blending of distinct polymers, an alternative technique for improving the polymer performance is the inclusion of solid particles by melt mixing the components [4].

In the recent past, there has been increasing interest in the blends of polyamides (PAs) and polyolefins. Most of the commercial PAs are based on the petroleum-based monomers. However, an advancing interest in the development as well as commercialization of PAs based on bio-based monomers has been noted during the past few decades [5-11]. In this respect, PA1010 is a commercially developed engineering PA possessing properties like effective strength, toughness, abrasion resistance and elasticity [12-16].

The development of polymer composites with nanofiller systems provides enhanced thermal stability, modulus, electrical conductivity and tensile strength, along with low gas penetrability and flammability [17-20]. Graphene nanoplatelets (GNPs) are stacks of multi-layered graphene sheets [21,22] and have emerged as an effective nanofiller for polymer matrices. GNPs have a superior balance of cost and physical properties as compared to other fillers [23-26]. The GNP functionalization has been utilized as a successful technique to enhance the dispersion of particles and material properties [27,28]. Due to the nanometer sized laminated framework of GNP nanoplatelets, these can prevent the flow of heat as well as combustible gases and possess the ability for developing the flame retardancy and thermal properties of polymers. Nevertheless, the power-

Haleema Saleem and Vikas Mittal*, The Petroleum Institute (part of Khalifa University of Science and Technology), Abu Dhabi, UAE
*Current address: Bletchington, Wellington County, Australia

ful van der Waals interactions among the graphene sheets mean that
the GNPs have the tendency to agglomerate in the polymer matrices.
The uniform GNP dispersion in the polymers is very critical to
achieve an optimum advancement in properties [29].

The PA/polyolefin blends possess superior impact strength [30-
32], along with excellent thermal, mechanical as well as oil resistance properties [33-35]. In the recent past, Pötschke *et al.* [36],
Cao *et al.* [37] and Mohamadi *et al.* [38] examined the influence of
graphene on the polymer blend properties. Cao *et al.* [37] observed
that with the addition of 0.5 wt% graphene oxide sheets in immiscible polyamide/polyphenylene oxide (PA/PPO, 90/10) blend, the
droplet diameter of the dispersed minor phase (PPO) was significantly decreased by more than 1 order of magnitude, showing enhanced compatibility. Also, it was confirmed that the graphene oxide
sheets could also function as reinforcing fillers in the polymer
blends, thereby, causing a significant enhancement in the thermal
stability and mechanical performance.

In this study, the mechanical, thermal and structural characteristics of GNP filled PA1010/HDPE nanocomposites have been analyzed, so as to develop this important class of blend nanocomposites
further.

2.2 Experimental

2.2.1 Materials

HDPE grade PE BB2581 was supplied by Abu Dhabi Polymers Company Limited. It possessed a density of 0.958 g/cm^3 and melt flow of
0.40 g/10 min at a temperature 190 °C. Renewably sourced biodegradable polyamide PA1010 with the trade name Zytel® RS LC1200
BK385 (70 wt% renewably sourced ingredient) was provided by
DuPont. It possessed a density of 1.03 g/cm^3 and melt viscosity of
500 Pa.s at 220 °C. The graphene nanoplatelets with the product
name N002-PDR were purchased from Angstron Materials, USA. The
material possessed a density of ≤2.20 g/cm^3 and specific surface area in the range 400-800 m^2/g.

2.2.2 Preparation of PA1010/HDPE Blend

PA1010/HDPE blend (1:1 ratio) was prepared by the melt mixing
method. For this purpose, PA1010 (2 g) and HDPE (2 g) were added

to Thermo Scientific Haake Minilab II twin conical screw extruder. Mixing temperature of 220 °C was employed, and the mixing was performed for 5 min in co-rotating mode at 100 rpm, with 4 g batch size.

2.2.3 Preparation of PA1010/HDPE/GNP Nanocomposites

PA1010/HDPE/GNP nanocomposites were generated by the melt mixing method by utilizing appropriate quantity PA1010, HDPE and GNPs, using Thermo Scientific Haake Minilab II twin conical screw extruder. Mixing temperature of 220 °C was employed, and the mixing was performed for 5 min in co-rotating mode at 100 rpm, with 4 g batch size. A series of PA1010/HDPE/GNP nanocomposites were generated by varying the amount of GNPs as 1, 2, 3 and 5 wt%, as detailed in Table 2.1.

Table 2.1 Composition of the PA1010/HDPE blend nanocomposites

Blend/Nanocomposites	PA1010 (%)	HDPE (%)	GNP (%)
Pure PA1010 + HDPE blend	50.0	50.0	0.0
PA1010/HDPE + 1% GNPs	49.5	49.5	1.0
PA1010/HDPE + 2% GNPs	49.0	49.0	2.0
PA1010/HDPE + 3% GNPs	48.5	48.5	3.0
PA1010/HDPE + 5% GNPs	47.5	47.5	5.0

Subsequent to extrusion, the injection molding of the samples was performed using Thermo Scientific Haake Minijet Pro injection molding machine. The mold and cylinder temperatures were fixed as 125 °C and 235 °C, respectively. Further, the injection and post pressure were set as 570 and 470 bars, respectively, for 10 s. Different specimen shapes, like dumbbell and circular discs, were generated for each composition. The nanocomposites were designated as PAPE (PA1010/HDPE blend), PAPEGNP1 (PA1010/HDPE + 1 wt% GNPs), PAPEGNP2 (PA1010/HDPE + 2 wt% GNPs), PAPEGNP3 (PA1010/HDPE + 3 wt% GNPs) and PAPEGNP5 (PA1010/HDPE + 5 wt% GNPs).

2.2.4 Characterization

The tensile properties of the PA1010/HDPE blend and PA1010/HDPE blend nanocomposites were analyzed using Instron

5567 Universal Testing Machine (UTM) at 10 mm/min strain rate. The tests were performed in accordance with ASTM D 638 at room temperature. The dumbbell-shaped specimens of 4 mm width, 53 mm length and 2 mm thickness were utilized. Win Test Analysis software was used to examine the magnitude of the mechanical properties, and an average of five specimens was recorded.

Calorimetric behavior of PAPE blend and PA1010/HDPE/GNP nanocomposites was examined utilizing differential scanning calorimetry (DSC) under N_2 atmosphere. Utilizing the TRIOS software, the DSC thermograms were attained using TA Discovery DSC. The extruded samples of about 3-8 mg weight were used for the characterization. DSC analysis was carried out under the non-isothermal conditions. Two heating and cooling cycles were performed. During the first cycle, the specimens were heated from 25 °C to 275 °C and subsequently cooled to -30 °C at a rate of +/- 10 °C/min in order to analyze the re-crystallization temperature. Subsequently, during the second cycle, the samples were re-heated to 275 °C in order to record the melt temperature, and later cooled to -30 °C at a rate of +/- 10 °C/min. Overall, peak melting temperature (T_m), peak crystallization temperature (T_C), enthalpy of melting (ΔH_m) and enthalpy of crystallization (ΔH_C) were examined.

Attenuated total reflection-Fourier transform infrared spectroscopy (ATR-FTIR) analysis of the pure GNPs, pristine PA1010, PA1010/HDPE blend and PA1010/HDPE/GNP nanocomposites was performed on the injection molded specimens utilizing Bruker VERTEX 70 spectrometer, equipped with a diamond ATR accessory, DLaTGS detector with KBr window and KBr beam-splitter. The ATR accessory possessed a ZnSe crystal having a refractive index of about 2.4@1000 cm^{-1} and a depth penetration of almost 2 μm at 45° angle of incidence. IR spectra in the 4000-600 cm^{-1} range were plotted employing the OPUS software at 8 cm^{-1} resolution with 64 average scans.

Panalytical X'Pert Pro diffractometer was employed for performing the wide angle X-ray diffraction (WAXRD) of pristine PA1010, HDPE, pure GNPs, PA1010/HDPE blend and PA1010/HDPE/GNP nanocomposites, using the scanning angle range of $2\theta = 5\text{-}40°$ and step size of 0.020° s^{-1}. The analysis was carried out in reflection mode, and the power settings of the instrument were 40 kV and 40 mA.

The thermal conductivity of the specimens was analyzed utilizing laser flash analysis (LFA) (LFA 447 from Netzsch, operated with the help of Nanoflash software). Thermal conductivity (λ) was calculated employing the relation $\lambda = \alpha.\rho.Cp$ where α, ρ and Cp are thermal diffusivity, density and specific heat of the specimens, respectively.

The Raman spectroscopic analysis was performed employing confocal microRaman spectrometer Horiba LabRAM HR, equipped with 514 nm laser source.

2.3 Results and Discussion

The tensile properties of the materials are presented in Table 2.2. The incorporation of GNPs was observed to significantly enhance the tensile modulus of the PA1010/HDPE blend (Figure 2.1). For

Table 2.2 Tensile properties of PA1010/HDPE blend and PA1010/HDPE/GNP nanocomposites

Material	Elongation at break (%)	Tensile stress at yield (MPa)	Tensile Modulus (MPa)
Pure PA1010	19.8	50	1084
PAPE blend	7.2	28	1254
PAPEGNP1	4.3	25	1309
PAPEGNP2	2.1	19	1450
PAPEGNP3	1.2	12	1612
PAPEGNP5	1.1	11	1639

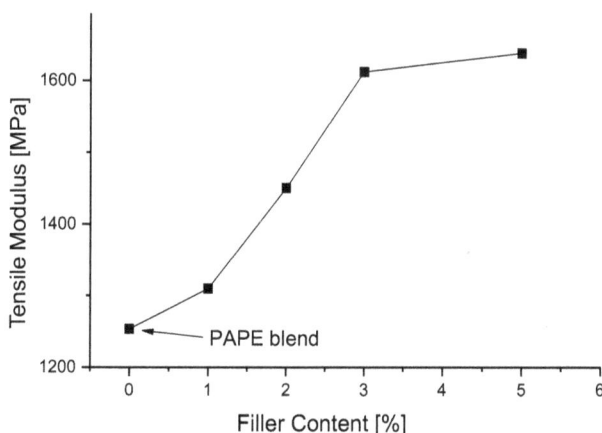

Figure 2.1 Tensile modulus of PA1010/HDPE blend and PA1010/HDPE/GNP nanocomposites.

instance, the PAPE blend exhibited the tensile modulus of 1254 MPa, which was enhanced to 1638 MPa for 5% GNPs reinforced PA1010/HDPE, thereby, displaying an increase of 31%. The pristine PA1010 exhibited a tensile strength of 50 MPa. The strength was observed to significantly reduce in the PA1010/HDPE blend, as compared to pristine PA1010. Incorporation of GNPs gradually re-duced the tensile strength further to 24.48 MPa for 1% GNPs addi-tion and 10.61 MPa for 5% GNPs inclusion. Further, it was con-firmed that the PA1010/HDPE blend exhibited a gradual reduction in the elongation at break, as a function of GNP concentration.

Calorimetric properties of the materials are outlined in Table 2.3. Figures 2.2 and 2.3 also demonstrate the melting and cooling curves

Table 2.3 Calorimetric behavior of PA1010/HDPE blend and PA1010/HDPE/GNP nanocomposites

Composition	Heating			Cooling		
	ΔH_m (cal/g)	ΔH_m (J/g)	T_m (°C)	ΔH_c (cal/g)	ΔH_c (J/g)	T_c (°C)
Pure PA1010	6.3	26.4	191	10.6	44.2	177
PAPE blend	1.4	5.8	199	4.2	17.5	178
PAPEGNP1	1.4	5.7	198	4.0	16.8	182
PAPEGNP2	3.8	15.9	199	4.7	19.7	182
PAPEGNP3	2.9	12.3	199	4.5	18.8	183
PAPEGNP5	3.9	16.3	198	5.3	22.2	183

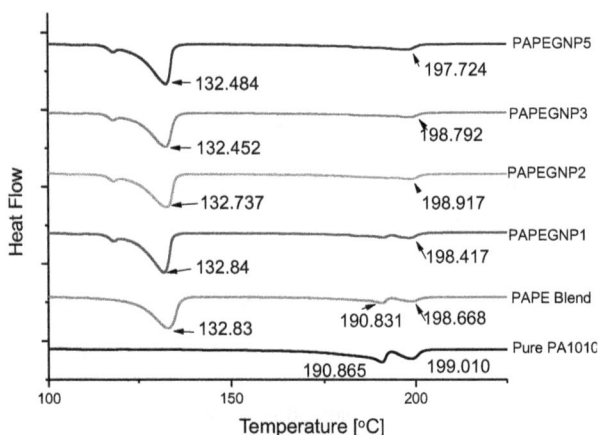

Figure 2.2 DSC melting curves of PA1010/HDPE blend and PA1010/HDPE/GNP nanocomposites.

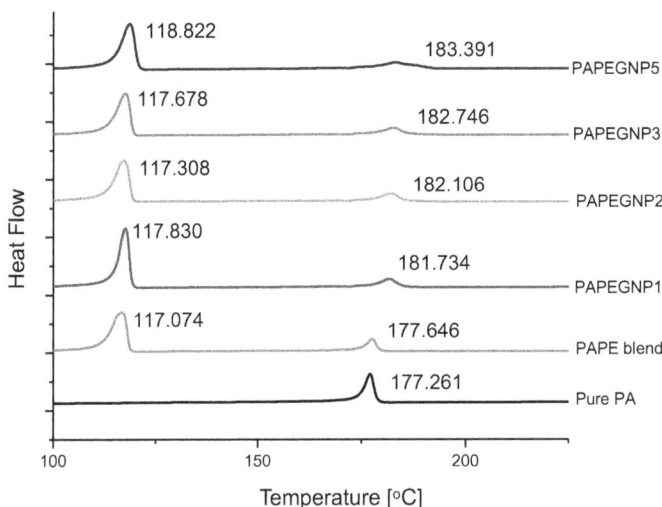

Figure 2.3 DSC cooling curves of PA1010/HDPE blend and PA1010/HDPE/GNP nanocomposites.

of the samples. PA1010 displayed two melting peaks at 191 °C and 199 °C. The observed multiple melting behavior may be due to the re-crystallization at the time of the DSC analysis. The peak situated at 191 °C might have resulted due to the melting of unstable crystals, whereas the peak at 199 °C can be attributed to the melting of highly stable crystals [39]. On incorporation of GNP nanoplatelets, the peak melting temperature of the PA phase in the composites displayed an unaltered value of 198 °C. A slight increment in the ΔH_m values (PA phase) was noted for the composites, relative to the PA1010/HDPE blend. It indicated that the presence of GNP nanofiller in the blend moderately promoted the polymer crystallization. The crystallization temperature of the PA1010/HDPE blend also improved marginally on incorporating GNPs, with the maximum value for the nanocomposite with 5% GNPs.

Pure PA1010 displayed the carbonyl stretching vibrations (amide I) of hydrogen bonded carbonyl groups at 1636 cm^{-1} and the N-H stretching modes (amide A) of hydrogen bonded N-H groups at 3305 cm^{-1} (Figure 2.4) [40,41]. Coupling of C-N stretching as well as N-H in-plane bending modes (amide II) was noted at 1538 cm^{-1}. It was also found that the carbonyl stretching vibrations (amide I) of hydrogen bonded carbonyl groups, the peak seen at 1636 cm^{-1}, in the PA1010/HDPE/GNP nanocomposites displayed a reduction in the peak intensity in the case of PAPEGNP1 and PAPEGNP2, as com-

pared to the PA1010/HDPE blend. However, for the PAPEGNP3 and
PAPEGNP5 nanocomposites, an increase in the intensity of the peak

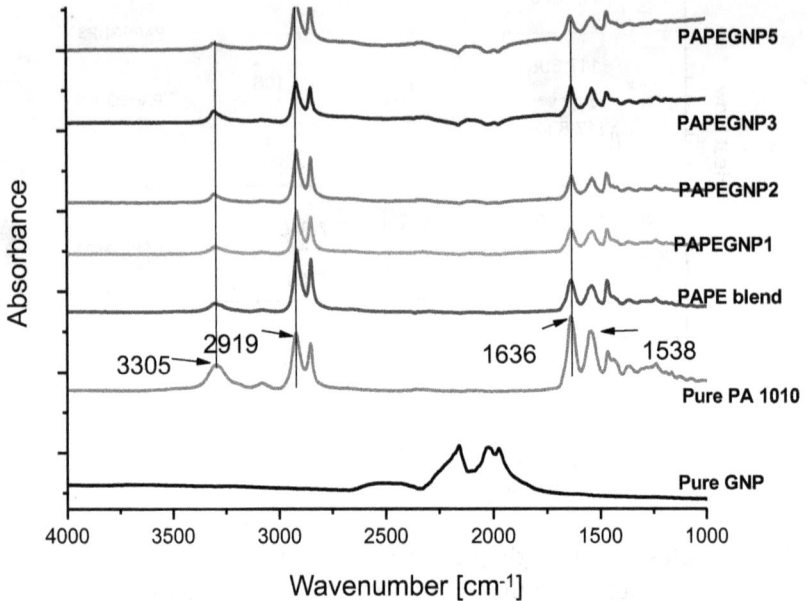

Figure 2.4 FTIR spectra of pristine GNPs, pure PA1010, PA1010/HDPE
blend and PA1010/HDPE/GNP nanocomposites.

at 1636 cm^{-1} was observed, relative to the PA1010/HDPE blend.
This was quantitatively examined by calculating the area ratio of
this band (A$_{1636}$) relative to the C-H stretching vibrations observed
at 2919 cm^{-1} (Table 2.4). A$_{1636}$/A$_{2919}$ for PA1010/HDPE blend was
observed to be lower as compared to the nanocomposites. It indicated that the particular amide bonds in the disordered amorphous
phase of the PA1010/HDPE blend underwent chemical interaction

Table 2.4 A$_{1636}$/A$_{2925}$ for pure PA1010, PA1010/HDPE blend and
PA1010/HDPE/GNP nanocomposites

Material	A$_{1636}$	A$_{2919}$	A$_{1636}$/A$_{2925}$
Pure PA1010	0.476	0.368	1.29
PAPE blend	0.211	0.402	0.52
PAPEGNP1	0.181	0.295	0.61
PAPEGNP2	0.194	0.359	0.54
PAPEGNP3	0.318	0.356	0.89
PAPEGNP5	0.323	0.463	0.70

in the presence of GNP nanoplatelets.

Figure 2.5 demonstrates the WAXRD patterns of the materials. Pristine PA1010 exhibited two distinct diffractions peaks at 2θ = 24.5° and 20.7°, corresponding to the (110/010) and 100 planes, respectively of the α-phase crystal [42,43]. Two major diffraction peaks were seen in the XRD pattern of HDPE: a strong reflection at 21.9° and a less intensive peak at 24.5° 2θ, corresponding to the orthorhombic unit cell framework of (110) and (200) reflection

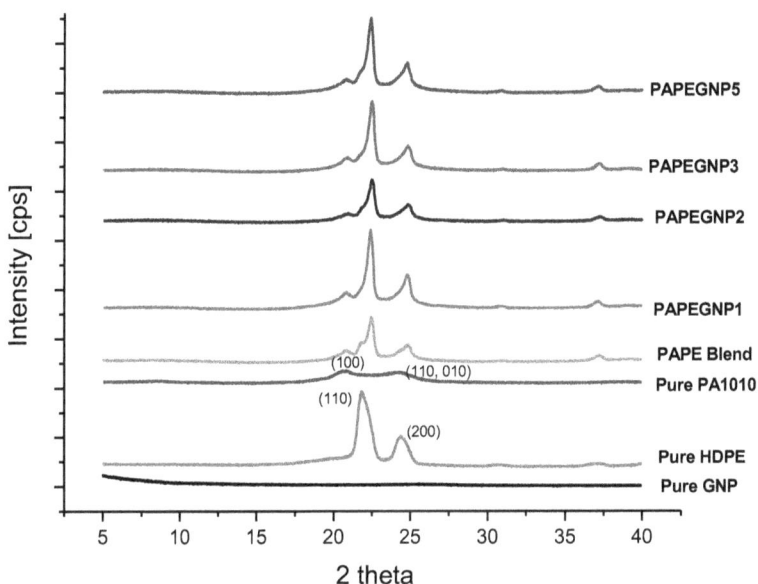

Figure 2.5 XRD patterns of pure PA1010, pure GNPs, PA1010/HDPE blend and PA1010/HDPE/GNP nanocomposites.

planes, respectively. For the PA1010/HDPE blend, the HDPE diffraction peak of (200) crystalline plane overlapped with the PA1010 diffraction peak of (110/010) at 2θ = 24.7°, and the HDPE diffraction peak of (110) crystal plane overlapped with the PA1010 diffraction peak of (100) at 2θ = 22.3°. For the PA1010/HDPE/GNP nanocomposites, the two diffraction peaks displayed higher values of 2θ, though there was no effect of the filler concentration on the peak positions. Further, the intensity of the diffraction peaks also showed a minor increment in the case of PA1010/HDPE/GNP nanocomposites, relative to the PA1010/HDPE blend. This indicated that the inclusion of GNP nanofiller to the PA1010/HDPE blend led to enhanced crystallinity.

The thermal conductivity of suspended graphene is almost 5000 W/m.K [44], thus, indicating the potential of conductivity improvement in polymers on the incorporating GNPs. The thermal conductivity of the blend and blend nanocomposites is depicted in Table 2.5 and Figure 2.6. The inclusion of GNPs was observed to enhance

Table 2.5 Thermal conductivity of PAPE blend and different PAPE nanocomposites

Temp.	PAPE Blend	PAPEGNP1	PAPEGNP2	PAPEGNP3	PAPEGNP5
20 °C	0.201	0.248	0.263	0.275	0.292
40 °C	0.181	0.219	0.230	0.244	0.260
60 °C	0.161	0.188	0.205	0.217	0.232

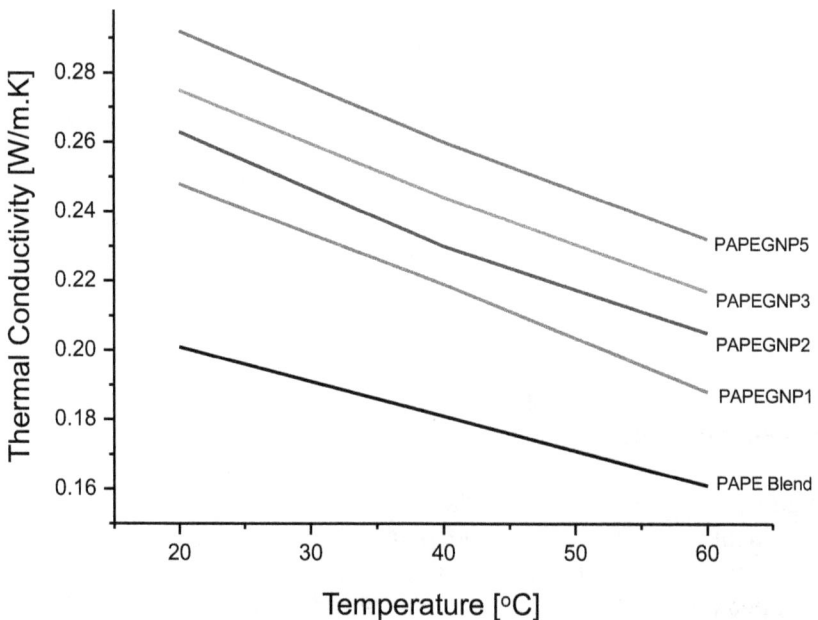

Figure 2.6 Thermal conductivity of PAPE blend and different PAPE nanocomposites.

the thermal conductivity of the PA1010/HDPE blend remarkably, as a function of filler concentration. At 60 °C, the nanocomposites with 3 wt% and 5 wt% GNP concentration exhibited the thermal conductivity values of 0.217 W/m.K and 0.232 W/m.K, respectively, indicating 35% and 44% increment, as compared to the HDPE/PA1010 blend.

Raman spectrum of pristine GNPs was collected in the range 1000-3000 cm^{-1}, as shown in Figure 2.7. The spectrum was observed to be similar to single crystal graphite [45]. The peak present at 1588 cm^{-1} was associated with the G band, whereas the peak at 1322 cm^{-1} corresponded to the D band. Also, the second-order Raman band (2D) was observed at 2656 cm^{-1}. HDPE spectrum presented typical Raman modes of PE (Figure 2.8). The bands noticed at approximately 1137 cm^{-1} and 1069 cm^{-1} were associated with the C-C stretching modes in the anisotropic and crystalline phases of the pure PE matrix. PA1010 spectrum comprised of the distinguishing

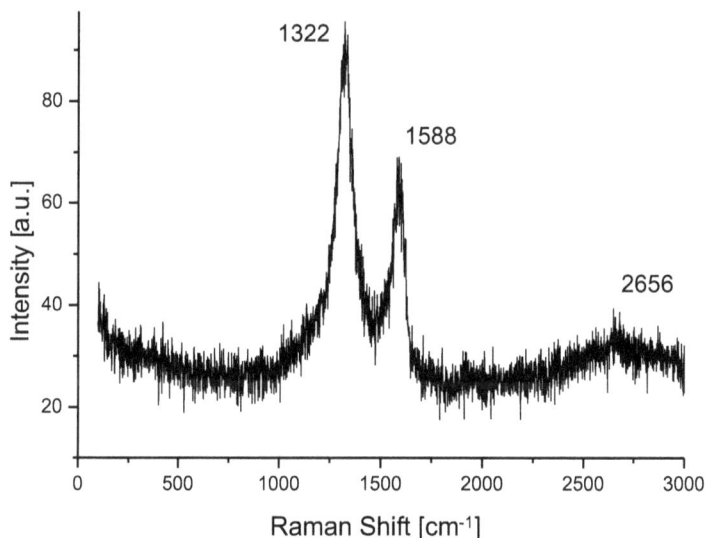

Figure 2.7 Raman spectrum of pure GNP.

bands of PA, in which the bands observed at 1304 cm^{-1} (amide III) and 1645 cm^{-1} (amide I) were related to the presence of the amide groups in the polymer matrix. The intensive bands observed at 2926 cm^{-1} and 2850 cm^{-1} corresponded to the asymmetric and symmetric CH$_2$ stretching vibrations, respectively. In the Raman spectra of the PA1010/HDPE/GNP nanocomposites (Figure 2.9), the bands were observed to be shifted to higher frequency, relative to the pristine GNPs. The aforementioned shift in the D and G bands confirmed the significant interaction between the GNPs and PA1010/HDPE blend. The analysis of the Raman I_D/I_G ratio (I_D – D-peak Raman intensity,

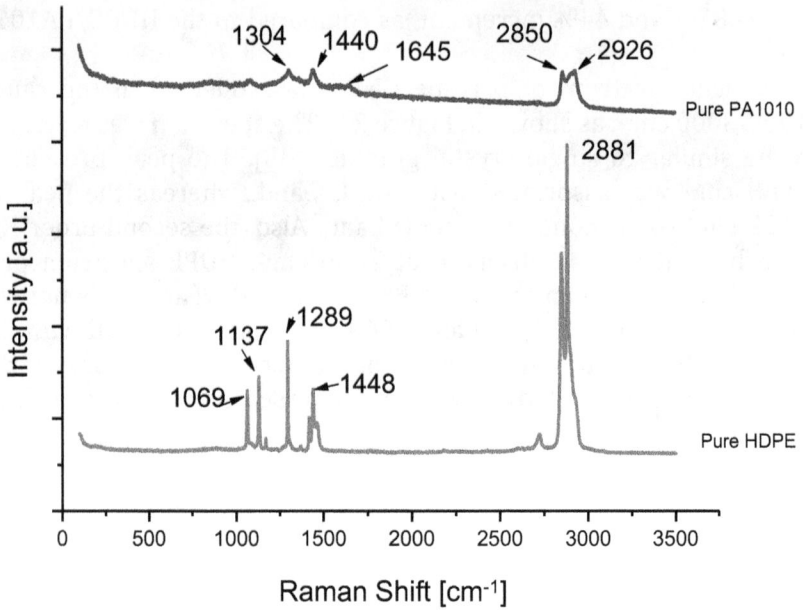

Figure 2.8 Raman spectra of pure HDPE and pure PA1010.

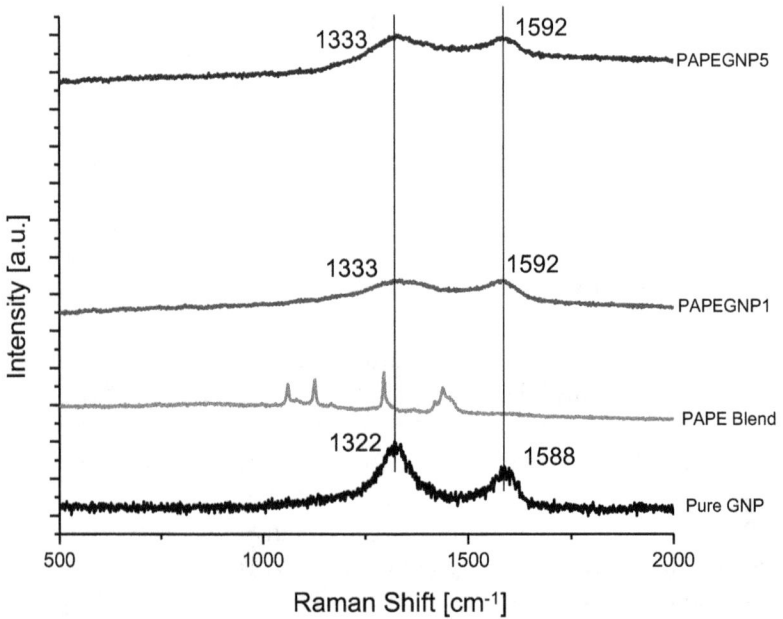

Figure 2.9 Raman spectra of PA1010/HDPE/GNP nanocomposites.

I_G – G-peak Raman intensity) of pure GNPs, PA1010/HDPE/1%GNP and PA1010/HDPE/5%GNP nanocomposites is detailed in Table 2.6. The value of I_D/I_G was greater in the case of PAPEGNP5 nanocomposite relative to PAPEGNP1. Higher I_D/I_G ratio indicated possible covalent interactions between the PA1010/HDPE blend and GNPs.

Table 2.6 Raman intensity ratio (I_D/I_G) of pristine GNPs and PA1010/HDPE nanocomposites

Material	D band (cm^{-1})	G band (cm^{-1})	I_D	I_G	Raman Intensity ratio (I_D/I_G)
Pure GNP	1322	1588	94	69	1.362
PAPEGNP1	1333	1592	3063	3055	1.002
PAPEGNP5	1333	1592	1591	1559	1.020

2.4 Conclusion

In this study, the mechanical, thermal properties and structural properties of polyamide1010 (PA1010)/high density polyethylene (HDPE) blend nanocomposites with graphene nanoplatelets (GNPs) have been evaluated. The incorporation of GNPs significantly enhanced the tensile modulus of the PA1010/HDPE blend, as a function of filler content. An increase of 31% in the modulus was observed for the composite with 5% filler content. The melting temperatures of the PA1010/HDPE blend and PA1010/HDPE/GNP nanocomposites did not indicate any significant change. A slight increment in the melt enthalpy was noted on GNPs addition, indicating enhanced crystallization in the composites. The A_{1636}/A_{2919} for the PA1010/HDPE blend was observed to be lesser as compared to the PA1010/HDPE/GNP composites. It indicated that the amide bonds in the disordered amorphous phase of the PA1010/HDPE blend underwent chemical interaction in the presence of GNP nanofiller. The XRD results also confirmed qualitatively the increased crystallinity on incorporating GNPs. Additionally, the presence of GNPs improved the thermal conductivity of the PA1010/HDPE blend, as a function of filler content. The Raman spectra of the PA1010/HDPE/GNP composites exhibited the shifting of the bands to higher frequency, relative to the pristine GNPs. The developed blend composites represent high value nanomaterials with a strong potential of use in a wide range of engineering applications.

References

1. Holsti-Miettinen, R., Seppälä, J., and Ikkala, O. T. (1992) Effects of compatibilizers on the properties of polyamide/polypropylene blends. *Polymer Engineering and Science*, **32**(13), 868-877.
2. La Mantia, F. P. (1993) Blends of polypropylene and nylon 6: Influence of the compatibilizer, molecular weight, and processing conditions. *Advances in Polymer Technology*, **12**(1), 47-59.
3. *Micro and Nanostructured Multiphase Polymer Blend Systems: Phase Morphology and Interfaces*, Harrats, C., Thomas, S., and Groeninckx, G. (eds.), CRC Press, USA (2005).
4. Jordan, J., Jacob, K. I., Tannenbaum, R., Sharaf, M. A., and Jasiuk, I. (2005) Experimental trends in polymer nanocomposites - A review. *Materials Science and Engineering A*, **393**(1), 1-11.
5. Bordes, P., Pollet, E., and Avérous, L. (2009) Nano-biocomposites: biodegradable polyester/nanoclay systems. *Progress in Polymer Science*, **34**(2), 125-155.
6. Carraher, Jr., C. E., Morrison, A., Roner, M. R., Moric-Johnson, A., Al-Huniti, M. H., and Miller, L. (2015) Metallocene-containing polyesters from reaction of 3, 5-pyridinedicarboxylic acid and metallocene dihalides and their preliminary ability to inhibit cancer cell growth. *Journal of the Chinese Advanced Materials Society*, **3**(4), 310-327.
7. Jia, Y., He, H., Peng, X., Meng, S., Chen, J., and Geng, Y. (2017) Preparation of a new filament based on polyamide-6 for three-dimensional printing. *Polymer Engineering and Science*, **57**(12), 1322–1328
8. Kausar, A. (2016) Polycarbonate/polypropylene-graft-maleic anhydride and nano-zeolite-based nanocomposite membrane: Mechanical and gas separation performance. *Advances in Materials Science*, **16**(4), 17-28.
9. Marchildon, K. (2011) Polyamides–still strong after seventy years. *Macromolecular Reaction Engineering*, **5**(1), 22-54.
10. Ma, L., Niu, H., Cai, J., Zhao, P., Wang, C., Lian, Y., Bai, X., and Wang, W. (2014) Optical, electrochemical, photoelectrochemical and electrochromic properties of polyamide/graphene oxide with various feed ratios of polyamide to graphite oxide. *Journal of Materials Chemistry C*, **2**(12), 2272-2282.
11. Zheng, D., Tang, G., Zhang, H. B., Yu, Z. Z., Yavari, F., Koratkar, N., Lim, S. H., and Lee, M. W. (2012) In situ thermal reduction of graphene oxide for high electrical conductivity and low percolation threshold in polyamide 6 nanocomposites. *Composites Science and Technology*, **72**(2), 284-289.
12. Zhishen, M., Qingbo, M., Jinhua, F., Hongfang, Z., and Donglin, C. (1993) Crystal structure and thermodynamic parameters of nylon-

1010. *Polymer International*, **32**(1), 53-60.

13. Zhang, S. L., Wang, G. B., Jiang, Z. H., Wang, D., Ma, R. T., and Wu, Z. W. (2005) Impact properties, phase structure, compatibility, and fracture morphology of polyamide-1010/thermoplastic poly (ester urethane) elastomer blends. *Journal of Polymer Science, Part B: Polymer Physics*, **43**(10), 1177-1185.

14. Lu, Z., Wu, A., Ou, X., Zhang, S., Niu, J., Ji, S., and Ling, Y. (2017) Enhanced anti-aging and mechanical properties of polyamide 1010 by sol-hydrothermal synthetic titanium dioxide-coated kaolinite addition. *Journal of Alloys and Compounds*, **693**, 381-388.

15. Gonzalez-Montiel, A., Keskkula, H., and Paul, D. R. (1995) Impact-modified nylon 6/polypropylene blends: 1. Morphology-property relationships. *Polymer*, **36**(24), 4587-4603.

16. Mittal, V. (2008) Mechanical and gas permeation properties of compatibilized polypropylene–layered silicate nanocomposites. *Journal of Applied Polymer Science*, **107**(2), 1350-1361.

17. Bréchet, Y., Cavaille, J. Y., Chabert, E., Chazeau, L., Dendievel, R., Flandin, L., and Gauthier, C. (2001) Polymer based nanocomposites: Effect of filler-filler and filler-matrix interactions. *Advanced Engineering Materials*, **3**(8), 571-577.

18. Favier, V., Cavaille, J. Y., Canova, G. R., and Shrivastava, S. C. (1997) Mechanical percolation in cellulose whisker nanocomposites. *Polymer Engineering and Science*, **37**(10), 1732-1739.

19. Carretero-González, J., Retsos, H., Verdejo, R., Toki, S., Hsiao, B. S., Giannelis, E. P., and López-Manchado, M. A. (2008) Effect of nanoclay on natural rubber microstructure. *Macromolecules*, **41**(18), 6763-6772.

20. Wang, K., Liang, S., Deng, J., Yang, H., Zhang, Q., Fu, Q., Dong, X., Wang, D., and Han, C. C. (2006) The role of clay network on macromolecular chain mobility and relaxation in isotactic polypropylene/organo-clay nanocomposites. *Polymer*, **47**(20), 7131-7144.

21. Kausar, A., Anwar, Z., Khan, L. A., and Muhammad, B. (2017) Functional graphene nanoplatelet reinforced epoxy resin and polystyrene-based block copolymer nanocomposite. *Fullerenes, Nanotubes and Carbon Nanostructures*, **25**(1), 47-57.

22. Kulkarni, H., Tambe, P., and Joshi, G. (2017) High concentration exfoliation of graphene in ethyl alcohol using block copolymer surfactant and its influence on properties of epoxy Nanocomposites. *Fullerenes, Nanotubes and Carbon Nanostructures*, **25**(4), 241-249.

23. Wang, F., Drzal, L. T., Qin, Y., and Huang, Z. (2016) Enhancement of fracture toughness, mechanical and thermal properties of rubber/epoxy composites by incorporation of graphene nanoplatelets. *Composites, Part A: Applied Science and Manufacturing*, **87**, 10-22.

24. Wang, F., Drzal, L. T., Qin, Y., and Huang, Z. (2015) Mechanical properties and thermal conductivity of graphene nanoplate-

let/epoxy composites. *Journal of Materials Science*, **50**(3), 1082-1093.

25. Ma, P. C., Siddiqui, N. A., Marom, G., and Kim, J. K. (2010) Dispersion and functionalization of carbon nanotubes for polymer-based nanocomposites: a review. *Composites, Part A: Applied Science and Manufacturing*, **41**(10), 1345-1367.

26. Cui, L. J., Wang, Y. B., Xiu, W. J., Wang, W. Y., Xu, L. H., Xu, X. B., Meng, Y., Li, L.Y., Gao, J., Chen, L. T., and Geng, H. Z. (2013) Effect of functionalization of multi-walled carbon nanotube on the curing behavior and mechanical property of multi-walled carbon nanotube/epoxy composites. *Materials and Design*, **49**, 279-284.

27. Anwar, Z., Kausar, A., Khan, L. A., and Muhammad, B. (2016) Modified graphene nanoplatelet and epoxy/block copolymer-based nanocomposite: physical characteristic and EMI shielding studies. *Nanocomposites*, **2**(3), 141-151.

28. Paszkiewicz, S., Szymczyk, A., Livanov, K., Wagner, H. D., and Rosłaniec, Z. (2015) Enhanced thermal and mechanical properties of poly (trimethylene terephthalate-block-poly (tetramethylene oxide) segmented copolymer based hybrid nanocomposites prepared by in situ polymerization via synergy effect between SWCNTs and graphene nanoplatelets. *Express Polymer Letters*, **9**(6), 509–524.

29. Terrones, M., Martín, O., González, M., Pozuelo, J., Serrano, B., Cabanelas, J. C., Vega-Díaz, S. M., and Baselga, J. (2011) Interphases in graphene polymer-based nanocomposites: Achievements and challenges. *Advanced Materials*, **23**(44), 5302-5310.

30. Utracki, L. A., Dumoulin, M. M., and Toma, P. (1986) Melt rheology of high density polyethylene/polyamide-6 blends. *Polymer Engineering and Science*, **26**(1), 34-44.

31. Chen, C. C., Fontan, E., Min, K., and White, J. L. (1988) An investigation of instability of phase morphology of blends of nylons with polyethylenes and polystyrenes and effects of compatibilizing agents. *Polymer Engineering and Science*, **28**(2), 69-80.

32. Hobbs, S. Y., Bopp, R. C., and Watkins, V. H. (1983) Toughened nylon resins. *Polymer Engineering and Science*, **23**(7), 380-389.

33. Willis, J. M., and Favis, B. D. (1988) Processing-morphology relationships of compatibilized polyolefin/polyamide blends. Part I: The effect of an ionomer compatibilizer on blend morphology. *Polymer Engineering and Science*, **28**(21), 1416-1426.

34. Koulouri, E. G., Georgaki, A. X., and Kallitsis, J. K. (1997) Reactive compatibilization of aliphatic polyamides with functionalized polyethylenes. *Polymer*, **38**(16), 4185-4192.

35. Holsti-Miettinen, R. M., Perttilä, K. P., Seppälä, J. V., and Heino, M. T. (1995) Oxygen barrier properties of polypropylene/polyamide 6 blends. *Journal of Applied Polymer Science*, **58**(9), 1551-1560.

36. Pötschke, P., Göldel, A., Gültner, M., and Liebscher, M. (2014) Localization Behavior of Carbon nanotubes and Graphene Nanoplatelets in Melt-Mixed Immiscible Polymer Blends, *POLYCHAR 22 World Forum on Advanced Materials*, South Africa.

37. Cao, Y., Zhang, J., Feng, J., and Wu, P. (2011) Compatibilization of immiscible polymer blends using graphene oxide sheets. *ACS Nano*, **5**(7), 5920-5927.

38. Mohamadi, S., and Sharifi-Sanjani, N. (2011) Investigation of the crystalline structure of PVDF in PVDF/PMMA/graphene polymer blend nanocomposites. *Polymer Composites*, **32**(9), 1451-1460.

39. Mittal, V., Chaudhry, A. U., and Luckachan, G. E. (2014) Biopolymer–thermally reduced graphene nanocomposites: structural characterization and properties. *Materials Chemistry and Physics*, **147**(1), 319-332.

40. Yoshioka, Y., Tashiro, K., and Ramesh, C. (2003) Structural change in the Brill transition of Nylon m/n (2) conformational disordering as viewed from the temperature-dependent infrared spectral measurements. *Polymer*, **44**(20), 6407-6417

41. Yoshioka, Y., and Tashiro, K. (2003) Structural change in the Brill transition of Nylon m/n (1) Nylon 10/10 and its model compounds. *Polymer*, **44**(22), 7007-7019.

42. Xu, Z., and Gao, C. (2010) In situ polymerization approach to graphene-reinforced nylon-6 composites. *Macromolecules*, **43**(16), 6716-6723.

43. Li, Y., and Yan, D. (2001) Preparation, characterization and crystalline transition behaviors of polyamide 4 14. *Polymer*, **42**(11), 5055-5058.

44. Lehman, J. H., Terrones, M., Mansfield, E., Hurst, K. E., and Meunier, V. (2011) Evaluating the characteristics of multiwall carbon nanotubes. *Carbon*, **49**(8), 2581-2602.

45. Tuinstra, F., and Koenig, J. L. (1970) Raman spectrum of graphite. *The Journal of Chemical Physics*, **53**(3), 1126-1130.

3

A Novel Method for the Synthesis of Aerogels using Vacuum Drying

3.1 Introduction

The very first aerogel was produced by Kistler [1] in 1931. This startling discovery in materials science led to the development of a multitude of organic materials like nitrocellulose, gelatin or rubber, which can all be obtained by replacing the lyogel using a supercritical fluid. In spite of the interesting properties of the silicon based aerogels, the industrial applications are very limited due to the technical difficulties in the drying procedure, as supercritical drying of methanol is not very cost effective [2,3].

The application of the sol-gel chemistry to silica aerogel synthesis was achieved by late 1970s, based on the acid catalyzed polymerization of alkoxysilanes. The immense progress in the field of organic polymer chemistry in the last few decades has prompted renewed interest in the aerogel research and application [4]. Thus, organic aerogels have been obtained from a multitude of organic precursors, such as resorcinol/formaldehyde, phenol/furfural and polyurethanes, by sol-gel processes and subsequent supercritical drying [4-7]. Organic aerogels can be further converted into carbon aerogels by subsequent pyrolysis at 600-2000 °C. The exceptionally multifaceted properties of all three classes of aerogels, viz., silica, organic and carbon aerogels, include high nanoscale porosity, large specific surface area, extremely low density, sound propagation, thermal expansion, heat transmission, large infrared optical adsorption coefficient and good electrical conductivity. Thus, these new materials are immensely attractive for many applications [8-13].

In this chapter, novel aerogels were prepared using nanofibrillated cellulose (NFC). N-Methylmorpholine-N-oxide (NMMO) was used as the solvent to achieve complete dissolution of cellulose, and vac-

Hariharan Krithivasan and Vikas Mittal***, The Petroleum Institute (part of Khalifa University of Science and Technology), Abu Dhabi, UAE
*Current address: University of Waterloo, Waterloo, Canada; **Current address: Bletchington, Wellington County, Australia

uum drying technique was used to obtain the final aerogel product. Specific properties of these aerogels were subsequently studied, which include density, porosity and pore size characteristics, along with mechanical properties. The prime reason for the preparation these aerogels was to optimize their characteristics in order to enable their coating with graphene. Afterwards, these aerogels can be converted into hydrophobic materials for effective oil spill clean-up applications.

3.2 Materials and Synthesis

Cellulose was obtained from Sigma-Aldrich, in the form of nano-fibrillated cellulose (NFC). The solvent used in the preparation of the aerocellulose was N-methylmorpholine-N-oxide (50 wt% in H_2O) and was procured from Sigma-Aldrich. Poly(propyl gallate) (PGA) was used as the stabilizer. Absolute analytical grade ethanol was used for the regeneration and solvent exchange procedure.

Solution with 7.5 wt% cellulose in NMMO-H_2O was prepared by dissolving the NFC in the solution, along with 0.05 g of PGA stabilizer (Table 3.1). The solution was stirred at room temperature for 60 min. Subsequently, the cellulose/NMMO solution was charged into a rotary evaporator. The rotary evaporator was used to control the temperature and pressure while maintaining constant stirring. The temperature was increased to 60 °C, and the solution was stirred for 2 h to achieve complete dissolution. Afterwards, the pressure was reduced inside the rotation chamber, thus, resulting in the evaporation of the excess water in the solution, thereby leaving behind NMMO and cellulose mixture. The same procedure was followed for different concentrations of cellulose and NMMO. After removal of the excess water, the solution was poured in a casting mold for drying at room temperature for 72 h.

Table 3.1 Concentrations of materials for sample preparation

Material	Concentration (wt%)
NMMO (50 wt% in H_2O)	70-75
Cellulose (NFC)	7.5-12.5
Distilled water	10-20
PGA (stabilizer)	0.05 (g)

The regeneration was performed by transferring the solid cellulose/NMMO mold in analytical grade ethanol baths. The first step of regeneration was performed for 4-6 h. Afterwards, the second and third step of solvent exchange was performed by transferring the mold into fresh ethanol baths until NMMO was completely replaced by ethanol. The whole procedure was performed under room temperature and atmospheric pressure. The final stage of the aerogel preparation to replace ethanol by air was achieved by using vacuum dryer. The sample (cellulose/ethanol mold) was placed in the vacuum dryer and high vacuum was obtained (50 Psi). At these conditions, ethanol evaporated slowly without disturbing the cellulose structure. This formed the delicate cellulose aerogel structure, with ethanol totally replaced by air, thus, leaving pores in the structure of the final product. After 24 h, the sample was removed from the vacuum dryer, and the final aerogel was obtained. Figure 3.1 shows the three stages of aerogel preparation, whereas Figure 3.2 demonstrates the steps needed for the preparation of the aerogels.

Figure 3.1 Three stages of aerogel preparation.

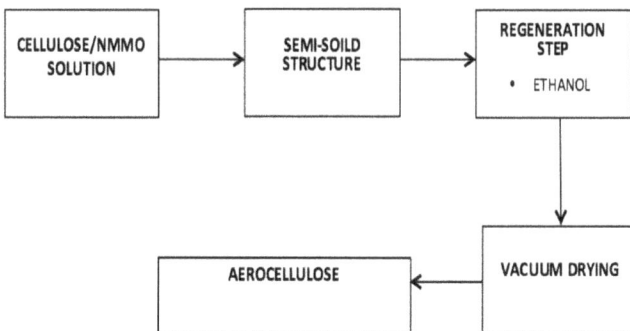

Figure 3.2 Aerogel preparation steps using vacuum dryer.

3.3 Results and Discussion

3.3.1 Density

The basic property to be characterized for the aerogels is their density. The density generally varies depending on various factors, e.g. the concentration (in wt%) of cellulose is the major contributor to the density of an aerogel. The number of regeneration baths used before the drying process also influence the density of the aerogels (Table 3.2). With the increase in the number of regeneration baths,

Table 3.2 Density and shrinkage measurement of aerogels at different cellulose concentrations

Cellulose Concentration	Number of Baths	Shrinkage (%)	Density (kg/m^3)
NFC (7.5 wt%)	2	32 %	153
	3	40%	
	4	52%	
NFC (8 wt%)	2	33%	168
	3	44%	
	4	57%	
NFC (8.5 wt%)	2	35%	177
	3	41%	
	4	52%	
NFC (9 wt%)	2	33%	192
	3	49%	
	4	58%	
NFC (9.5 wt%)	2	35%	197
	3	45%	
	4	59%	
NFC (10 wt%)	2	41%	210
	3	49%	
	4	62%	
NFC (11 wt%)	2	41%	223
	3	52%	
	4	68%	
NFC (12 wt%)	2	48%	241
	3	55%	
	4	62%	
NFC (12.5 wt%)	2	51%	251
	3	61%	
	4	69%	

the shrinkage factor of the aerogels increases. This leads to a much denser aerogel structure, as cellulose is packed in a tighter structure. Also, the drying time seems to have an impact on the density of the aerogels, with longer drying time leading to a reduction in the density of the aerogels after threshold time. Before threshold time, some ethanol is still left in the aerogel structure.

Overall, it can be seen from Table 3.2 that an increase in the concentration of cellulose enhanced the density of the aerogels. Also, increasing the number of regeneration baths resulted in enhanced shrinkage.

3.3.2 Thermal Conductivity

The thermal conductivity of the monolithic aerogels can be simply described as the addition of the solid thermal conductivity, gaseous thermal conductivity and radiative thermal conductivity. For this, the measurements were performed using a hot wire device. The platinum wire was squeezed between a pair of identical aerogels. The effective embedding of the wires was possible due to the elasticity of the aerogels. The specimens were enclosed inside a glass tube to reduce the noise from the environment. The platinum wire served as the heating element. The temperature increase in the aerogel during a heat pulse was recorded. It was clear that the aerogel density affected their thermal conductivity (Table 3.3). Cellulose is a poor conductor of heat, so the major heat conductor was the gas available

Table 3.3 Measurement of thermal conductivity for different cellulose concentrations

Cellulose Concentration	Thermal Conductivity $Wm^{-1}K^{-1}$
NFC (7.5 wt%)	0.05209
NFC (8 wt%)	0.05202
NFC (8.5 wt%)	0.05134
NFC (9 wt%)	0.05120
NFC (9.5 wt%)	0.05114
NFC (10 wt%)	0.05115
NFC (10.5 wt%)	0.05091
NFC (11 wt%)	0.04997
NFC (11.5 wt%)	0.04988
NFC (12 wt%)	0.04984
NFC (12.5 wt%)	0.04977

in the pores. When there is an increase in the density, the gas present inside the aerogels reduces, thus, leading to a reduction in the thermal conductivity.

3.3.3 Mechanical Properties

The stress-strain behaviors of the aerogels were studied by applying compressive stress to further understand the mechanical characteristics. The samples were compressed from 15 mm to 8 mm in thickness without any decrease in the mechanical load, as usually practiced for the stress tests in foamed plastics [14]. The compressive stress-strain curves were studied to evaluate the Young's modulus, compressive yield stress and yield strain (Table 3.4). From the data,

Table 3.4 Mechanical properties of cellulose based aerogels

Cellulose Concentration	$E\ (Nmm^{-2})$	$\sigma_y(Nmm^{-2})$	$\varepsilon_y(\%)$
NFC (7.5 wt%)	12.8	0.16	4.14
NFC (8 wt%)	13.7	0.18	4.11
NFC (8.5 wt%)	15.1	0.19	4.07
NFC (9 wt%)	16.3	0.22	4.09
NFC (9.5 wt%)	65.8	0.91	2.87
NFC (10 wt%)	66.7	0.96	2.81
NFC (10.5 wt%)	68.3	0.99	2.79
NFC (11 wt%)	68.9	1.02	2.75
NFC (11.5 wt%)	69.7	1.07	2.73
NFC (12 wt%)	70.4	1.09	2.77
NFC (12.5 wt%)	70.6	1.13	2.70

it can be seen that the initial cellulose concentration affected the mechanical properties of the aerogel significantly. The mechanical properties were poor until the cellulose concentration reached the threshold value (9.5 wt%). After this point, the mechanical properties of the aerogels improved significantly. The observed behavior resulted due to the relation between the pore characteristics and mechanical behavior. Before the threshold point, the pore structure collapsed, and the aerogel was damaged. After the threshold value was obtained, the aerogel did not collapse under compressive stress. Figure 3.3 also shows the scanning electron microscopy (SEM) image of the aerogel after tensile test indicating collapsed aerogel structure.

Figure 3.3 SEM image of the aerogel after tensile test.

3.3.4 Pore Characteristics

The pore characteristics of the aerogels were analyzed using nitrogen adsorption test, performed under constant temperature. The surface area, pore diameter and pore volume were obtained from the nitrogen adsorption test (Table 3.5). The isotherms for all aerogel samples were of type IV, thus, the material could be classified as

Table 3.5 Pore characteristics of cellulose based aerogels

Cellulose Concentration	BET Surface Area $(m^2 g^{-1})$	Pore Diameter (nm)	Pore Volume $(cm^3 g^{-1})$
NFC (7.5 wt%)	225	21.2	0.84
NFC (8 wt%)	230	22.6	0.87
NFC (8.5 wt%)	240	23.3	0.91
NFC (9 wt%)	245	25.2	0.93
NFC (9.5 wt%)	260	28.9	0.97
NFC (10 wt%)	270	30.8	0.96
NFC (10.5 wt%)	275	32.5	0.99
NFC (11 wt%)	280	34.1	1.03
NFC (11.5 wt%)	280	35.3	1.02
NFC (12 wt%)	290	36.4	1.06
NFC (12.5 wt%)	290	37.1	1.09

mesoporous [15]. SEM images also confirmed that the structure of the cellulose based aerogels majorly consisted of mesopores (Figure 3.4). The average mesopore diameter of the materials was observed in the range of 21 to 37 nm.

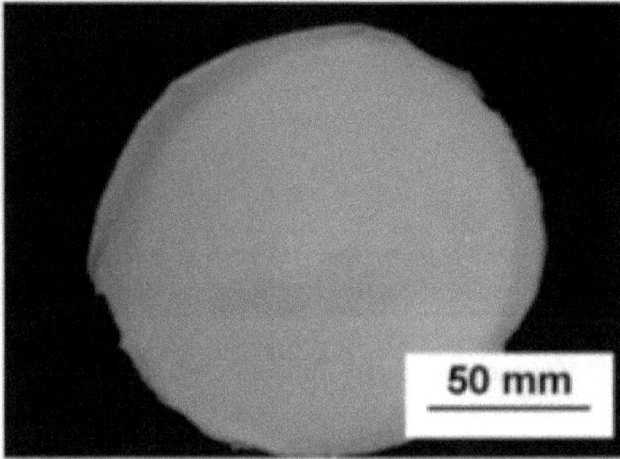

(a)

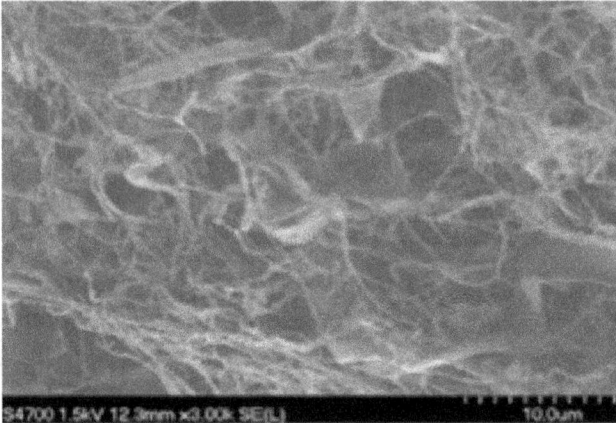

(b)

Figure 3.4 SEM image of the (a) aerogel and (b) pores in the aerogel.

3.4 Conclusion

Cellulose based aerogels were successfully synthesized using the vacuum drying process by employing different concentrations of nanofibrillated cellulose. The density of the aerogels ranged from 153 to 251 kg/m³, whereas the pore diameter was between 21 to 37 nm. The average thermal conductivity of the samples was observed to be 0.05 $Wm^{-1}K^{-1}$, and the porosity percentage ranged from 75 to 85%. The observed results confirmed that the vacuum drying can be effectively used to prepare aerogels, thus, making the process much simpler and cost-effective as compared the conventional techniques requiring the use of supercritical dryers. Further, these aerogels are intended to be coated with functionalized graphene to improve the mechanical properties and make the material hydro-phobic for oil spill clean-up applications.

References

1. Kistler, S. S. (1931) Coherent expanded aerogels and jellies. *Nature*, **127**, 741.
2. Chen, W., Yu, H., Li, Q., Liu, Y., and Li, J. (2011) Ultralight and highly flexible aerogels with long cellulose in nanofibers. *Soft Matter*, **7**(21), 10360-10368.
3. Chen, N., and Pan, Q. (2013) Versatile fabrication of ultralight magnetic foams and application for oil–water separation. *ACS Nano*, **7**(8), 6875-6883.
4. Pekala, R. (1989) Organic aerogels from the polycondensation of resorcinol with formaldehyde. *Journal of Materials Science*, **24**(9), 3221-3227.
5. Phalippou, J. Woignier, T. and Prassas, M. (1990) Glasses from aerogels. *Journal of Materials Science*, **25**(7), 3111-3117.
6. Biesmans, G., Mertens, A., Duffours, L., Woignier, T., and Phalippou, J. (1998) Polyurethane based organic aerogels and their transformation into carbon aerogels. *Journal of Non-Crystalline Solids*, **225**, 64- 68.
7. Reynolds, G., Fung, A., Wang, Z., Dresselhaus, M., and Pekala, R. (1995) The effects of external conditions on the internal structure of carbon aerogels. *Journal of Non-Crystalline Solids*, **188**(1), 27-33.
8. Husing, N., and Schubert, U. (2006) Aerogels. In: *Ullmann's Encyclopedia of Industrial Chemistry*, Wiley-VCH, Germany, doi:10.1002/14356007.c01_c01.
9. Bag, S., Arachchige, I. U., and Kanatzidis, M. G. (2008) Aerogels from metal chalcogenides and their emerging unique properties. *Journal of Materials Chemistry*, **18**(31), 3628-3632.
10. Farmer, J. C., Fix, D. V., Mack, G. V., Pekala, R. W., and Poco, J. F. (1996) Capacitive deionization of NaCl and NaNO₃ solutions with carbon aerogel electrodes. *Journal of the Electrochemical Society*,

143(1), 159-169.

11. Pierre, A. C., and Pajonk, G. M. (2002) Chemistry of aerogels and their applications. *Chemical Reviews*, **102**(11), 4243-4266.

12. Akimov, Y. K. (2003) Fields of application of aerogels (review). *Instruments and Experimental Techniques*, **46**(3), 287-299.

13. Emmerling, A., and Fricke, J. (1992) Small angle scattering and the structure of aerogels. *Journal of Non-Crystalline Solids*, **145**, 113-120.

14. Liebner, F., Haimer, E., Potthast, A., Loidl, D., Tschegg, S., Neouze, M.-A., Wendland, M., and Rosenau, T. (2009) Cellulosic aerogels as ultra-lightweight materials. Part 2: Synthesis and properties. *Holzforschung*, **63**(1), 3-11.

15. Rouquerol, F., Rouquerol, J., and Sing, K. (1999) *Adsorption by Powders and Porous Solids*, Academic Press, USA.

4

Cobalt Ferrite Doped Graphene/Polyacrylamide Cryogel Nanocomposites: Synthesis and Characterization

4.1 Introduction

Cryogels have large and well-defined interconnected porous structure. These materials are heterogeneous and non-superimposable about few micrometer to few hundred micrometers [1,2]. Free-radical cryo-polymerization technique has been utilized for preparing cryogels from polymeric gel precursors below the solvent freezing point, where the function of the freeze-drying process is to maintain the interconnected structure of the resultant cryogel [3]. The properties of cryogels such as pore dimensions, pore wall thickness, mechanical toughness and elasticity depend on the conditions of preparation, amount of solvent and precursors, temperature gradients, cooling rate and other conditions [4,5]. High porosity is exhibited by the cryogels, thus, these can be utilized as an adsorbent at the macroscopic level for wastewater treatment or oil spill clean-up.

In this study, graphene-based carbonaceous monolithic material containing cobalt ferrite nanoparticles has been synthesized for adsorption applications. During the formation of polyacrylamide based cryogels, divalent metal ions (such as Co^{2+} and Fe^{2+}) perform as both crosslinker and catalyst due to their binding ability towards oxygen groups [6] which leads to crosslinking between graphene oxide and polyacrylamide skeletons, thus, enhancing the strength of the whole cryogel.

4.2 Materials and Methods

4.2.1 Materials

Natural graphite powder (10 mesh), H_2SO_4 (95-97%, Merck), H_3PO_4 (85%, Merck), $KMnO_4$ (Eurolab), HCl (37%, Panreac), hydrogen per-

Naman Arora and Vikas Mittal*, The Petroleum Institute (part of Khalifa University of Science and Technology), Abu Dhabi, UAE
*Current address: Bletchington, Wellington County, Australia

oxide (35%, Merck), $CoCl_2.6H_2O$ (Sigma-Aldrich), iron(III) nitrate na-
nohydrate (≥98%, Sigma-Aldrich), acrylamide, polyacrylamide
(PAAm), tetramethylethylenediamine (TEMED), N,N'-meth-
ylenebis(acrylamide) (MBA) and potassium persulphate (KPS) were
used in the study.

4.2.2 Preparation of Graphene Oxide (GO)

Graphene oxide (GO) was prepared by using the improved Tour's
method [7]. Graphite (10 g) was added to a 9:1 molar mixture of
H_2SO_4 (540 mL) and H_3PO_4 (60 mL). The mixture was stirred for 30
min, followed by the gradual addition of $KMnO_4$ (56 g). The stirring
was continued for further 72 h. The reaction was continued by slowly
pouring H_2O_2 and subsequently allowing it to stir for 8 h until the mixture color turned yellow. Produced graphene oxide was washed with
1 M HCl solution until no sulfate ions were detected, followed by
washing with excess distilled water to remove chloride ions. The GO
solution was then poured into Teflon petri dishes for freeze drying.

4.2.3 Synthesis of Graphene based Polyacrylamide Cryogel containing Cobalt Ferrite Nanoparticles

GO exfoliation was carried out under ultra-sonication (2 h) by adding
GO (0.1 g) in deionized water (50 mL). Later, $CoCl_2.6H_2O$ (0.6 g) and
$Fe(NO_3)_3.9H_2O$ (0.5 g) were added as the Co^{2+} and Fe^{2+} sources, followed by stirring for 2 h. Polyacrylamide (PAAm) (10 g) and N,N'-
methylenebis(acrylamide) (2 g) were subsequently added to the solution, followed by continuous stirring for 4 h to form a homogeneous
solution. Finally, 100 μL TEMED and 0.2 g KPS were added as initiators. Hydrogel was obtained after 7 min, followed by washing with
excess distilled water to remove the residual ions. Subsequently, the
hydrogel was subjected to lyophilization process for 5 d at a temperature of -89 °C and vacuum of 10 Pa to obtain a GO based polyacrylamide cryogel. The cryogel was reduced by dipping into hydrazine hy-
drate solution for 2 h at 75 °C with slow agitation. Finally, graphene
based polyacrylamide cryogel containing cobalt ferrite nanoparticles
($CoFe_2O_4$/graphene/PAAm) was obtained.

4.2.4 Characterization of Cryogel

Wide-angle X-ray diffraction (WAXRD) pattern of the sample was collected by employing analytical powder (X' Pert PRO) diffractometer, using CuKα (1.5406 Å wavelength) radiation in reflection mode. The sample was step-scanned between 5 and 60° at a step size of 2θ = 0.02° s^{-1}.

Infrared spectra of the sample was recorded by attenuated total reflectance Fourier transform infrared spectroscopy (FTIR-ATR), using a Bruker Vertex 70 spectrometer.

Scanning electron microscopy (SEM) and energy dispersive X-ray (EDX) analysis were carried out to ascertain the morphology of the sample, using FEI Quanta FEG250 equipment at accelerating voltages of 10-20 kV.

Transmission electron microscopy (TEM) analysis was also carried out to obtain the morphology of $CoFe_2O_4$/graphene/PAAm nanocomposite. The sample for TEM was prepared by dispersing 0.5 mg of adsorbent in 20 mL of dimethylformamide under sonication at room temperature for 10 min. Two drops of suspension were poured onto a 400 mesh copper electron microscopy grid covered with thin holey carbon film, followed by drying in air.

4.3 Results and Discussion

Figure 4.1a shows the digital image of the graphene oxide-polyacrylamide based cryogel containing divalent metal salts of Co^{2+} and Fe^{2+} (GO-PAAm). After carbonization process, the GO-PAAM based cryogel turned into black carbonaceous nanocomposite cryogel ($CoFe_2O_4$/graphene/PAAm), as shown in Figure 4.1b. To mention again, these cryogels have been synthesized by obtaining the initial suspensions as hydrogels and further lyophilization of the corresponding hydrogels in order to eliminate the water present in the structure, which results in dried monoliths/cryogels with desirable shapes, depending on the glass molds. It can be noticed that GO-PAAm was a homogeneous monolith with yellow-brownish color. After the reduction of GO-PAAM, divalent metal ions (Co^{2+} and Fe^{2+}) and GO are converted into cobalt ferrite nanoparticles and graphene, respectively. As a result, $CoFe_2O_4$/graphene/PAAm carbonaceous monolith was obtained.

WAXRD pattern of the resultant cryogel is represented in Figure 4.2. The observed diffraction peaks corresponded to the cubic $CoFe_2O_4$ spinel phase without any detectable impurities [8]. It indicated the effective synthesis of the desired cryogel structure. The SEM micrograph of $CoFe_2O_4$/graphene/PAAm is also represented in

(a)

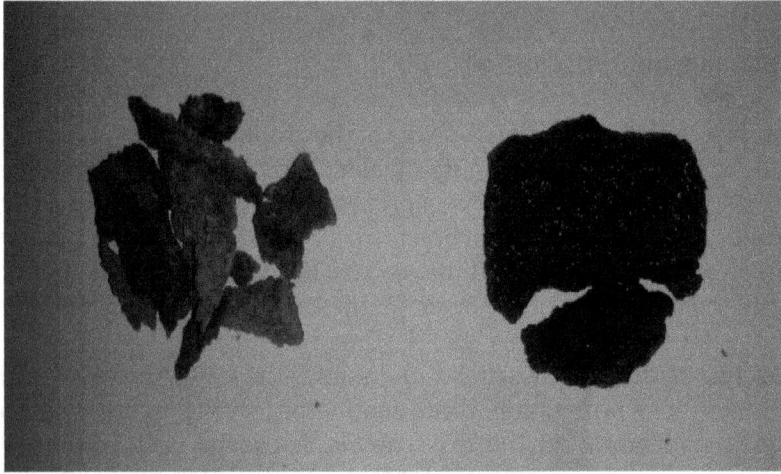

(b)

Figure 4.1 (a) Graphene oxide based polyacrylamide (GO-PAAm) cryogel and (b) reduction of GO-PAAm cryogel to form $CoFe_2O_4$/graphene/PAAm cryogel.

Figure 4.3. Cobalt ferrite nanoparticles were observed to be present in a well-defined interconnected porous structure. Fourier transform infrared FTIR spectrum of the cryogel is also demonstrated in Figure 4.4. The amide groups in the cryogel were observed to exhibit two

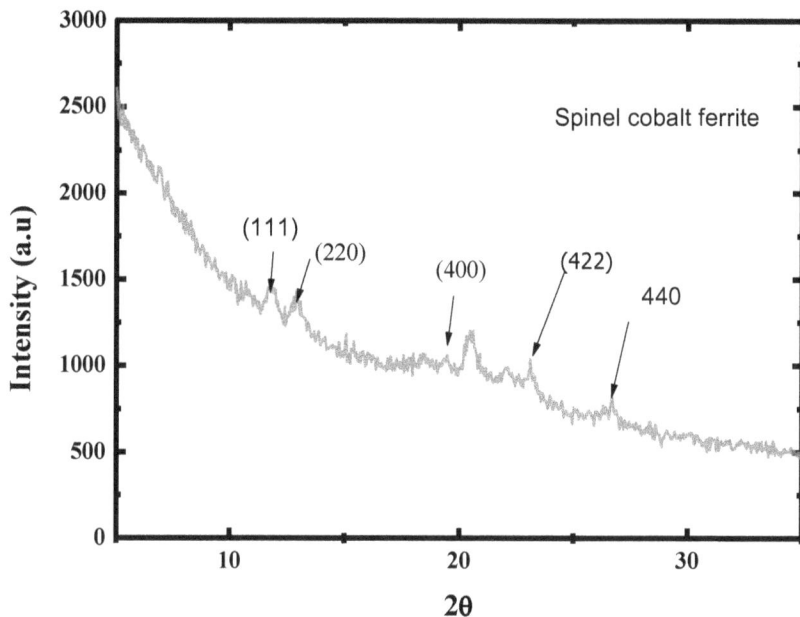

Figure 4.2 XRD spectrum of $CoFe_2O_4$/graphene/PAAm cryogel.

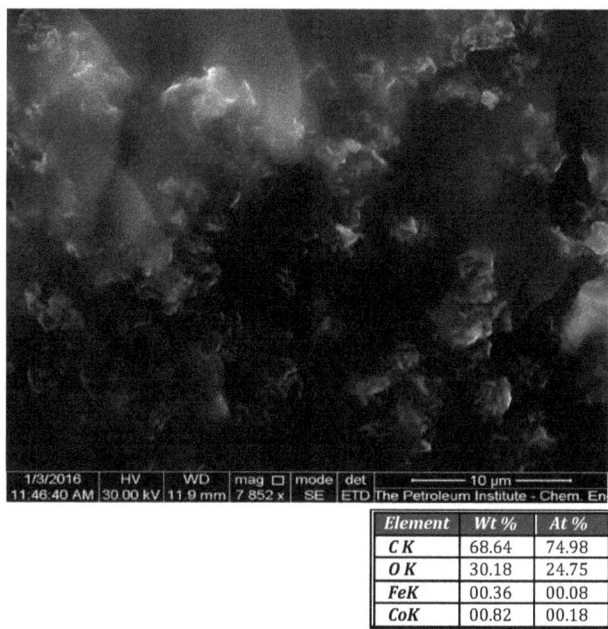

Element	Wt %	At %
C K	68.64	74.98
O K	30.18	24.75
FeK	00.36	00.08
CoK	00.82	00.18

Figure 4.3 SEM image and EDX analysis for the cryogel.

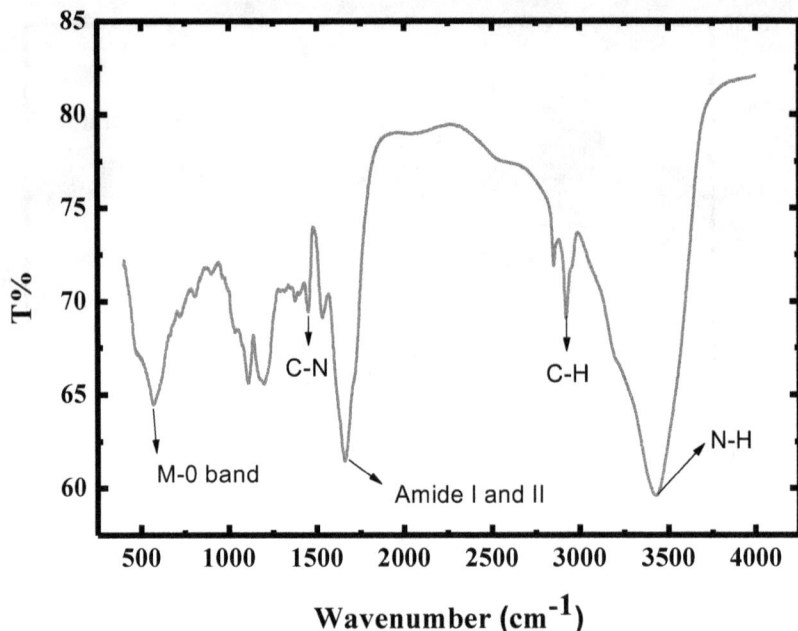

Figure 4.4 FTIR spectra of CoFe$_2$O$_4$/graphene/PAAm cryogel.

characteristic peaks, the first peak at 1672 cm^{-1} (amide I and II) representing the –C=O stretching (amide I) and –NH$_2$ bending vibrations of the amide groups (amide II), whereas the second peak at 1425 cm^{-1} was attributed to the –CN stretching vibration of the amide groups [9]. The peaks at 601 cm^{-1} and 1142 cm^{-1} were attributed to the v-ferrite structure (M-O) and Fe-Co alloys system, respectively [10].

TEM image of the resultant adsorbent are presented in Figure 4.5a. It can be observed that the carbonaceous matrix contained graphene sheets, thus, attributing enhanced surface area to the adsorbent. Additionally, cobalt ferrite nanoparticles were uniformly distributed onto graphene sheets, which corresponded to additional active adsorption sites and subsequently increased interactions with negatively charged contaminated ions. Uniform porosity could also be noticed in Figure 4.5b, thus, confirming that resultant adsorbent was a porous material.

4.4 Conclusion and Future Work

In this communication, CoFe$_2$O$_4$/graphene/PAAm nanocomposite cryogel has been synthesized by employing *in-situ* polymerization.

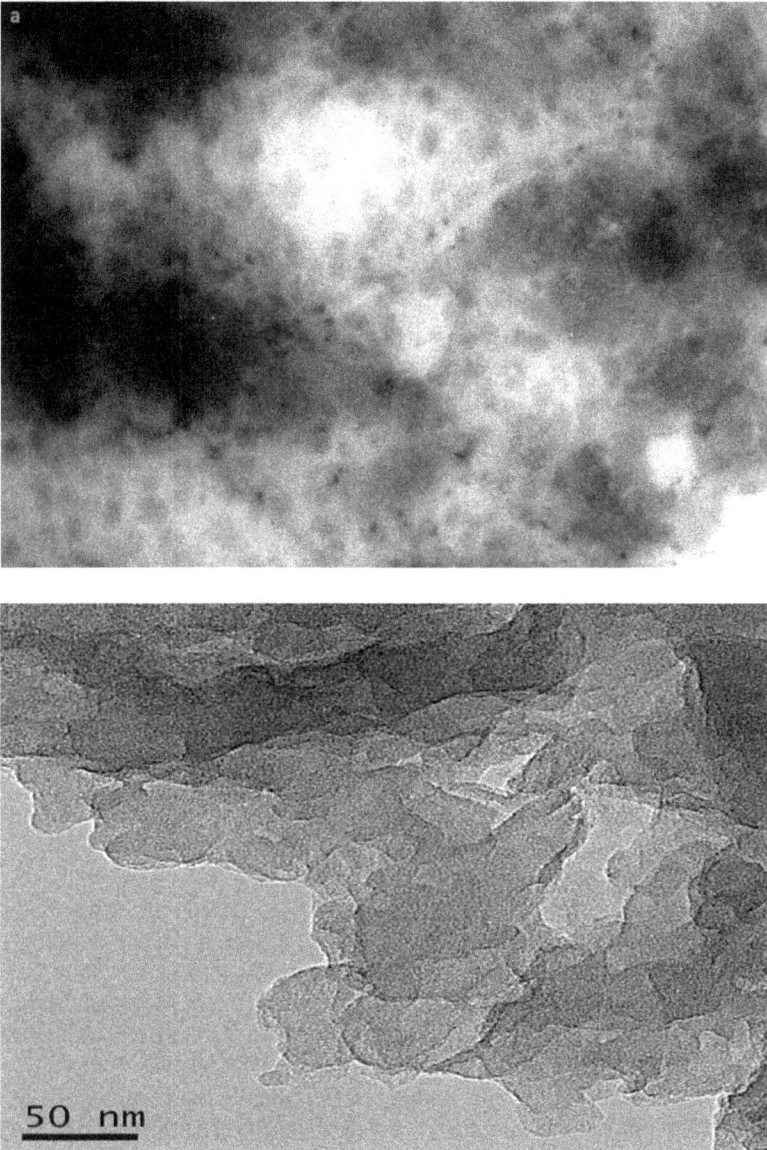

Figure 4.5 TEM images of CoFe$_2$O$_4$/graphene/PAAm cryogel.

The cryogel exhibited interconnected porous structure which enables the use of the developed material as promising adsorbent for the removal of contaminants from aqueous solutions, especially industrial water effluents. The monolithic structure also presents advantages in terms of efficiency and handling.

References

1. Bibi, N. S., Singh, N. K., Dsouza, R. N., Aasim, M., and Fernández-La-hore, M. (2013) Synthesis and performance of megaporous immo-bilized metal-ion affinity cryogels for recombinant protein capture and purification. *Journal of Chromatography A*, **1272**, 145-149.
2. Zheng, S., Wang, T., Liu, D., Liu, X., Wang, C., and Tong, Z. (2013) Fast deswelling and highly extensible poly (N-isopropylacrylamide)-hectorite clay nanocomposite cryogels prepared by freezing polymerization. *Polymer*, **54**, 1846-1852.
3. Sahiner, N., Seven, F., and Al-lohedan, H. (2015) Superporous cryo-gel-M (Cu, Ni, and Co) composites in catalytic reduction of toxic phe-nolic compounds and dyes from wastewaters. *Water, Air and Soil Pollution*, **226**, 1-12.
4. Reichelt, S., Becher, J., Weisser, J., Prager, A., Decker, U., Moeller, S., Berg, A., and Schnabelrauch, M. (2014) Biocompatible polysaccha-ride-based cryogels. *Materials Science and Engineering C*, **35**, 164-170.
5. Petrov, P., Petrova, E., Tchorbanov, B., and Tsvetanov, C. B. (2007) Synthesis of biodegradable hydroxyethylcellulose cryogels by UV irradiation. *Polymer*, **48**, 4943-4949.
6. Park, S., Lee, K.-S., Bozoklu, G., Cai, W., Nguyen, S. T., and Ruoff, R. S. (2008) Graphene oxide papers modified by divalent ions—enhanc-ing mechanical properties via chemical cross-linking. *ACS Nano*, **2**, 572-578.
7. Marcano, D. C., Kosynkin, D. V., Berlin, J. M., Sinitskii, A., Sun, Z., Slesarev, A., Alemany, L. B., Lu, W., and Tour, J. M. (2010) Improved synthesis of graphene oxide. *ACS Nano*, **4**, 4806-4814.
8. Wang, W., Yang, H., Xian, T., and Jiang, J. (2012) XPS and magnetic properties of $CoFe2O4$ nanoparticles synthesized by a polyacryla-mide gel route. *Materials Transactions*, **53**, 1586-1589.
9. Chiem, L. T., Huynh, L., Ralston, J., and Beattie, D. A. (2006) An in situ ATR–FTIR study of polyacrylamide adsorption at the talc surface. *Journal of Colloid and Interface Science*, **297**, 54-61.
10. Rana, S., Philip, J., and Raj, B. (2010) Micelle based synthesis of co-balt ferrite nanoparticles and its characterization using Fourier transform infrared transmission spectrometry and thermogravim-etry. *Materials Chemistry and Physics*, **124**, 264-269.

5

PEBAX-MOF Membranes

5.1 Introduction

Polymeric membranes are micro-porous films that work as semi-permeable barriers for the separation of different gases. Polymer membranes based gas separation technology has various advantages, including low energy consumption, mechanical simplicity, easy scale up and smaller footprint [1]. An additional advantage of the polymeric membrane technology is the possibility to control the density, size, size distribution, shape and vertical alignment of membrane pores which is not possible in other types of membranes technologies [2]. However, polymer membranes suffer from a trade-off between permeability and selectivity, i.e. the polymers with high gas permeability generally have low gas selectivity and vice versa [1,3]. In order to overcome this issue, several inorganic fillers have been incorporated in the organic polymer matrices to generate mixed matrix membranes (MMMs), which possess merits of easy processability, superior permeability and high selectivity. Recently, metal organic frameworks (MOFs) have been considered as outstanding candidates to be used as fillers for MMMs. Several MOFs, such as MIL-53, MIL-101, ZIF-7, ZIF-8, ZIF-90, UiO-66, UiO-67, CuBTC and CuTPA have been identified for this purpose. Among various MOFs, HKUST-1 is a preferred material for the development of MMMs due to its three-dimensional morphology with interconnected porous network, which is beneficial for effective gas separation.

Among the various polymer matrices used to develop membranes, PEBAX-4533 thermoplastic elastomer has been widely employed. It is a block copolymer composed of covalently bonded hard segments of polyamide (PA), which provide mechanical support to the polymer, and soft segments of polyether (PE) with high free volume, which offer transport channels for effective gas permeation

Sara Alkhoori, M. R. Vengatesan, G. Karanikolos and Vikas Mittal, The Petroleum Institute (part of Khalifa University of Science and Technology), Abu Dhabi, UAE*
Current address: Bletchington, Wellington County, Australia

[4,5]. Different PA:PE ratios result in different PEBAX grades. PEBAX polymers have wide range of applications as they combine properties of both thermoplastics and elastomers.

Previous works have reported MMMs developed using different PEBAX grades containing altered PE and PA block lengths, along with various additives and nanofillers. Sheth *et al.* [1] studied four different series of PEBAX containing nylon 12 and poly(tetramethylene oxide). PEBAX with higher PA content was observed to exhibit enhanced Young's modulus. However, a decrease in elongation at break and ultimate tensile strength were accompanied due to the rigid PA chains. Thermal analysis indicated an enhancement in thermal stability of PEBAX with higher PA content, and a shift of glass transition temperature to higher values was observed. Moreover, PEBAX with higher PA content showed two melting and crystallization peaks. A study by Murali *et al.* [6] investigated the effect of multi-walled carbon nanotubes (MWCNTs) (from 0 to 5 wt%) and crosslinking on the permeation behavior of PEBAX-1675 (60% PE and 40% PA). Incorporation of MWCNTs increased the inter-segmental spacing of the polymer. Gas permeation analysis revealed that the filler improved the permeation of the polymer, specifically toward CO_2. The effect of different graphene fillers on the poly(vinyl alcohol) (PVA) membranes was also studied by Ovid'ko [3]. The author observed that graphene reinforcement was superior than carbon nanotubes due high modulus and 2D geometry. Mittal *et al.* [7] also carried out a similar study by using both carbon nanotubes and graphene as fillers for polymeric membranes. The authors underlined three key factors of filler dispersion, interaction between the filler and polymer matrix and alignment of the filler in the matrix. It was observed that using graphene nanoplatelets (GNPs) provided better mechanical properties than CNTs due to 2D morphology. Also, the high surface to volume ratio in GNPs allowed better stress transfer in the polymer matrix during loading.

Mixed matrix membranes of PEBAX-1657 loaded with 4A zeolite for gaseous separations were also developed by Murali *et al.* [8], and the main objective was to overcome the trade-off relationship limitations in polymeric membranes by utilizing the properties of organic polymer (good permeability) and inorganic zeolite (high selectivity due to small pore size and thermal stability). Other studies have also mentioned zeolites to be excellent fillers for separation purposes due to their molecular sieve ability and high thermal, mechanical and chemical stability [4]. A good dispersion of zeolite filler was observed for lower filler fractions. Also, corrugations were observed in the cross-sectional images which corresponded to the formation of void morphology due to the placement of the filler particles in between the polymer layers. Diffraction analysis exhibited an improvement in the amorphous nature (permeability) of the polymer, along with im-proved crystallinity of zeolite (which was indicated by sharper peaks), with increasing zeolite loading. Polymer membranes filled with 10 wt% zeolite showed good permeability and selectivity. More-over, permeability of CO_2 was observed to improve by increasing the pressure.

Different MOF based MMMs have also been developed in recent years [9,10]. The incorporation of HKUST-1 in the polymeric mem-branes has been observed to result in improved properties. To fur-ther explore the usefulness of MOFs for developing high performance membranes, MMMs involving PEBAX and HKUST-1 (in varying weight fractions) were developed and analyzed for their mechanical and structural characteristics in the current study.

5.2 Experimental

5.2.1 Materials

PEBAX-4533 was provided by ARKEMA in granular form. HKUST-1 MOF was synthesized at the Petroleum Institute laboratories by fol-lowing the literature reported protocols. The synthesized HKUST-1 was a fine blue powder with a purity of 99%. The solvents used in the study were acquired from Sigma Aldrich.

5.2.2 Optimization of PEBAX-4533 Dissolution and Membrane Preparation

The formation of fine membranes is directly related to the effective dissolution of PEBAX. The dissolution of PEBAX-4533 was studied us-ing different solvents such as NMP/ethanol, dimethyl sulfoxide (DMSO) and dimethylformamide (DMF). Around 10 mL of the solvent was used to dissolve 1 g of PEBAX-4533. In order to achieve a homo-geneous solution, it was observed that the polymer granules should be stirred in the solvent at 80 °C for 4 h. PEBAX was observed to dis-solve uniformly in most of the solvents. After several experiments, the PEBAX dissolution protocols were optimized with DMF as per the following steps:

- Stir the polymer solution at 80 °C for 4 h.
- Pre-heat the casting dish before casting the solution. This step is to prevent the formation of lumps within the membrane due to cooling down of the polymer solution during casting process.
- Solvent evaporation in a pre-heated oven at 85 °C for 24 h.

5.2.3 Preparation of PEBAX-4533/HKUST-1 based MMMs

MMM formulations were prepared by stirring a specific amount of HKUST-1 in 1 g of PEBAX-4533 in DMF at 80 °C for 4 h. Similar casting procedure, as used for pure PEBAX membranes, was employed for generating MMMs. MMMs with different filler fractions of 0, 0.25, 0.5, 1 and 3 wt% were generated.

5.2.4 Characterization

Tensile analysis was carried out using Instron 5567 Universal Testing Machine following ASTM D-638 standard. The dumbbell shaped samples of dimensions $17 * 3 * 0.1$ mm (length, width and thickness respectively) were analyzed at a strain rate of 10 mm/min. Five samples for each membrane were tested, and the mean was taken.

Calorimetric properties of the samples were analyzed using differential scanning calorimeter (DSC) from TA Instruments. 3-5 mg of the samples were used for the analysis. The heating cycle ranged from the ambient temperature up to 250 °C to remove previous processing history, followed by cooling to -50 °C, using a rate of 10°C/min under dry nitrogen atmosphere. The calorimetric data was taken from the second heating and cooling cycles.

Thermogravimetric analysis (TGA) was used to study the thermal stability of the polymer and MMMs using a TA discovery TGA employing TRIOS software. 3-5 mg of the samples were heated to 700 °C using a heating rate of 10 °C/min under inert gas conditions, and weight loss versus temperature curves were established.

Wide angle X-ray diffraction (WAXD) was utilized for the evaluation of the crystallinity of the polymeric membranes by applying Bragg's Law and the microstructure morphology (including the crystallites size) by the application of Scherrer's equation. The analysis was carried out on Panalytical X-Pert Pro Diffractometer using monochromatic CuKα radiation in reflection mode with filament current of 40 mA and accelerated voltage of 40 kV, by scanning between 5°

and 60° 2θ at room temperature. Zero background holder was used for lowering the noise.

Compositional analysis was performed at room temperature using attenuated total reflection Fourier transform infrared spectroscopy (ATR-FTR) between 4000–500 cm^{-1} on BRUKER Tensor II Platinum ATR with mid-infrared range detector.

Contact angle measurements were carried out using Rame-hartINC goniometer employing 5 µL deionized water (DIW) and 5 µL diiodomethane (DIM), both with known surface tension and contributions of polar and dispersive interactions (Table 5.1).

Table 5.1 Surface tension and corresponding contributions of polar and dispersive interactions

	Surface tension $(\frac{mJ}{m^2})$	Dispersive portion $(\frac{mJ}{m^2})$	Polar portion $(\frac{mJ}{m^2})$
DIM	50.8	50.8	0
DIW	72.8	21.8	51

5.3 Results and Discussion

Figures 5.1-5.3 illustrate the tensile results of the MMM samples. Pure

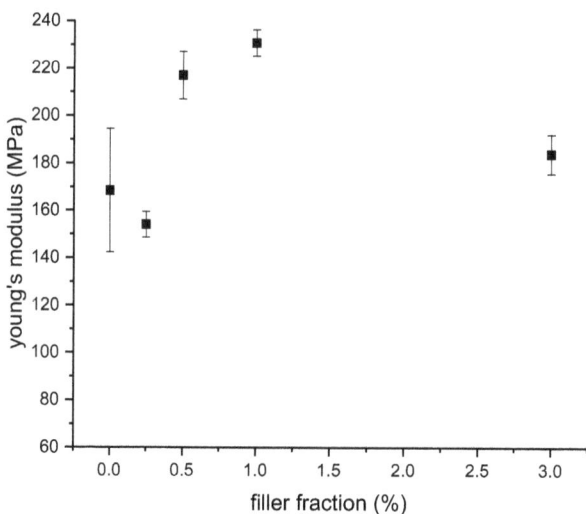

Figure 5.1 Tensile modulus of PEBAX and MMMs.

Polymers and Polymer Nanocomposites

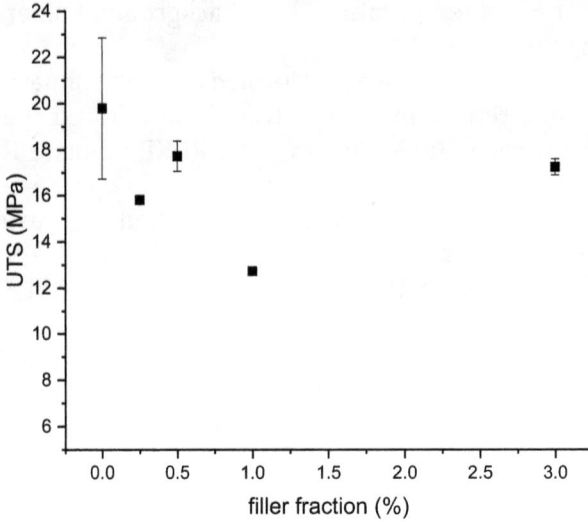

Figure 5.2 Ultimate tensile strength (UTS) of PEBAX and MMMs.

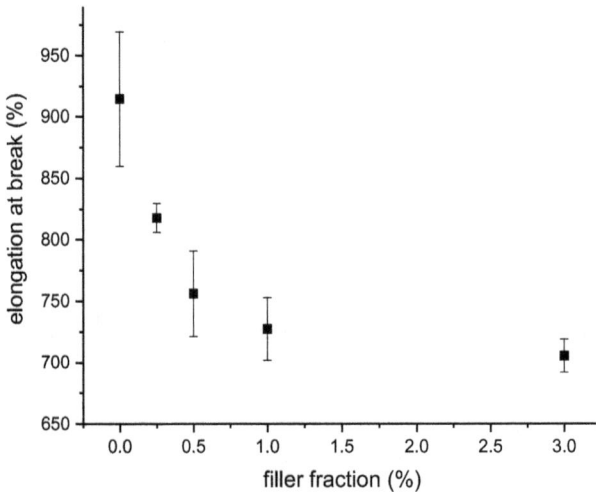

Figure 5.3 Elongation at break of PEBAX and MMMs.

PEBAX film was observed to have high elongation at break, moderate Young's modulus and low ultimate tensile strength (UTS), indicating a tough and non-rigid behavior similar to the soft PE segments. The incorporation of MOF nanofiller in PEBAX resulted in a decrease in both elongation at break and UTS, though the decrease in UTS was not significant. It indicated that the MOF filler acted as stress concentrator in the matrix resulting in a decrease in elongation at break, a common issue faced by glassy polymers due to their rigid structure [4,10-12]. The Young's modulus of the nanocomposite films was observed to enhance with filler fractions higher than 1%. This suggested a good degree of interaction and strass transfer between the filler and matrix. The high surface area of the filler allowed better stress transfer from the polymer matrix during loading. In contrast, it is interesting to note that the membrane with 0.25% loading resulted in a decrease in Young's modulus as compared to the pure polymer. This is due to the lower content of the MOF acting as a solid state plasticizer. Additionally, at a higher filler content of 3%, the modulus of the MMM exhibited a decrease due to a small degree of agglomeration of the filler in the polymer matrix, though the modulus was still higher than the pure polymer.

The melting peaks exhibited broad temperature range (Figure 5.4), around 50 °C, which is expected due to the wide range of crystallite size distribution [13]. PEBAX-4533 pellets exhibited one melting peak at 147 °C owing to prevalent thermal history (Table 5.2), whereas two melting peaks were observed for the PEBAX membrane. Also, in PEBAX samples with higher PA content, only one melting peak is generally observed in DSC, while the second melting peak at lower temperatures is related to the PE content of the copolymer in

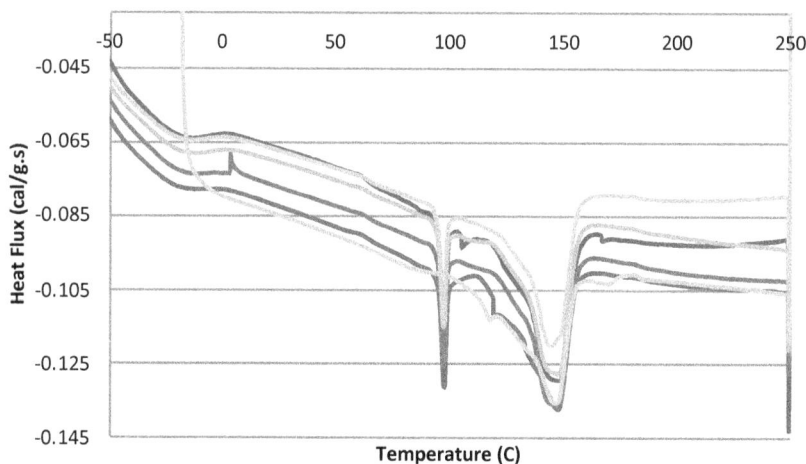

Figure 5.4 DSC second heating thermograms. The curves from bottom to top indicate PEBEX pellets, PEBAX film in DMF and nanocomposite films with 0.25, 0.5, 1 and 3% filler.

Table 5.2 DSC melting temperature and enthalpy of melting

Filler fraction (wt%)	PEBAX pellets	0	0.25%	0.5%	1%	3%
Lower bound T_m (°C)	None	98	98	98	98	98
Lower bound $\Delta H_{melting}$ $\left(\frac{J}{g}\right)$	None	1.8	2.5	2.4	2.4	2.5
Upper bound T_m (°C)	147	148	149	145	148	145
Upper bound $\Delta H_{melting}$ $\left(\frac{J}{g}\right)$	20	21	21	22	21	22

higher PE content grades [1]. At higher filler fractions, somewhat higher enthalpy values were observed probably due to the filler particles affecting the crystallinity of the resultant membranes.

Similar to the heating thermal scans, the cooling scans for the pure polymer pallets exhibited one crystallization dip at 75 °C, while all other samples had two crystallization dips at higher and lower temperature ranges due to the crystallization of PA segments at higher temperatures and PE segments at lower temperatures (Table 5.3).

Table 5.3 DSC crystallization temperature and enthalpy of crystallization

Filler fraction (wt%)	PEBAX pellets	0	0.25	0.5	2	3
Lower bound T_c (°C)	None	174	174	173	174	174
Lower bound $\Delta H_{crystallization}$ $\left(\frac{J}{g}\right)$	None	0.1	0.1	0.2	0.2	0.2
Upper bound T_c (°C)	75	93	110	111	117	111
Upper bound $\Delta H_{crystallization}$ $\left(\frac{J}{g}\right)$	22.2	21.0	23.1	23.2	23.4	23.0

The addition of MOF exhibited higher T_c at upper temperature range indicating fast crystallization in the MMMs. The enthalpy of crystallization was also observed to increase with increasing the filler fractions. The crystalline morphology of MOF particles acts as a nucleation center for the crystallization of the polymer chains, allowing them to arrange themselves into more perfect crystalline lamellae swiftly at higher temperature.

Also, the differences in the melting and crystallization behaviors of the PEBAX pellets and PEBAX film indicated that the orientation of crystals was more uniform in the film.

From the second heating thermograms, glass transition temperature (T_g) values of the MMMs were also recorded, as shown in Table 5.4. The T_g values of the MMMs increased on enhancing the filler content, which indicated a good degree of compatibility between the filler and polymer matrix.

Table 5.4 Glass transition temperature values of PEBAX and MMMs

Filler fraction (wt%)	$T_g(°C)$
0	-33.5
0.25	-31.8
0.5	-33.5
1	-32.9
3	-31.5

High thermal stability of HKUST-1 for high temperature applications has been confirmed by other studies. HKUST-1 degradation temperature varies based on the synthesis method. Some HKUST-1 degradation temperatures reported in literature are 220 °C [14], 240 °C [14,15], 300 °C [9,16], 310 °C [17], 350 °C [18] and 376 °C [19], thus, verifying the high thermal stability of the filler. The presence of impurities within the pore structure of MOF is indicated by the char yield or total weight loss from the 'true' weight of the solid. The TGA thermograms of the pure polymer and the MMMs exhibited a single degradation step. The materials exhibited the degradation in the temperature range 260-360 °C, with the thermal stability of the membranes slightly enhanced with the addition of filler. Table 5.5 summarizes the degradation temperatures of the membranes with different filler fractions. Table 5.6 also shows the char yield, taken at 700 °C, and the limited oxygen index (LOI) values calculated for the PEBAX and membrane samples.

Table 5.5 TGA degradation temperatures of membranes

	0%	0.25%	0.5%	1%	3%
5% weight loss	351	354	353	354	353
10% weight loss	363	367	366	368	366
50% weigh loss	402	404	404	406	406

Table 5.6 Char yield and the LOI of PEBAX and MMMs

Filler fraction (wt%)	0	0.25	0.5	1	3
Char yield %	1.48	1.32	1.86	2.09	2.00
LOI	18.1	18.0	18.2	18.3	18.3

XRD of HKUST-1 is presented in Figure 5.5, and it was in good agreement with the reported literature. HKUST-1 is highly crystalline material possessing face-centered cubic crystalline structure, and it shows several sharp narrow peaks with high intensities at 2θ angles

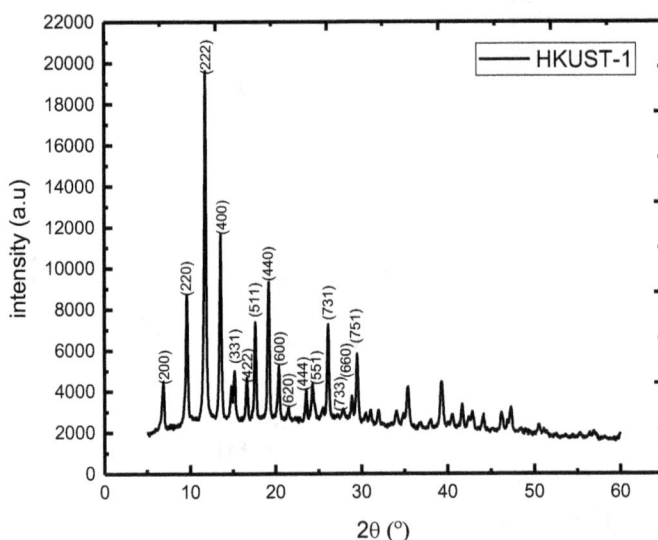

Figure 5.5 WAXRD pattern of HKUST-1.

around 7°, 12°, 14°, 18°, 19° and 26°, which evidence the microporous nature of the MOF. Moreover, the characteristic peak at around 9.6° 2θ corresponded to the *d*-spacing of 9.5 Å for HKUST-1's main pores [6,8,15,20]. Diffractograms of pure PEBAX-4533 and MMMs showed semi-crystalline nature, as shown in Figure 5.6. Pure PEBAX exhibited a narrow peak around $2\theta = 22.7°$. The MMMs exhibited similar XRD patterns as pure PEBAX, with a slight shifting of the peak to higher angles, indicating that the addition of MOF had no major effect on the crystalline structure of PEBAX. The absence of any characteristic peak of the filler particles in the composite membranes also indicated a uniform dispersion of the filler in the polymer matrix, irrespective of the filler concentration.

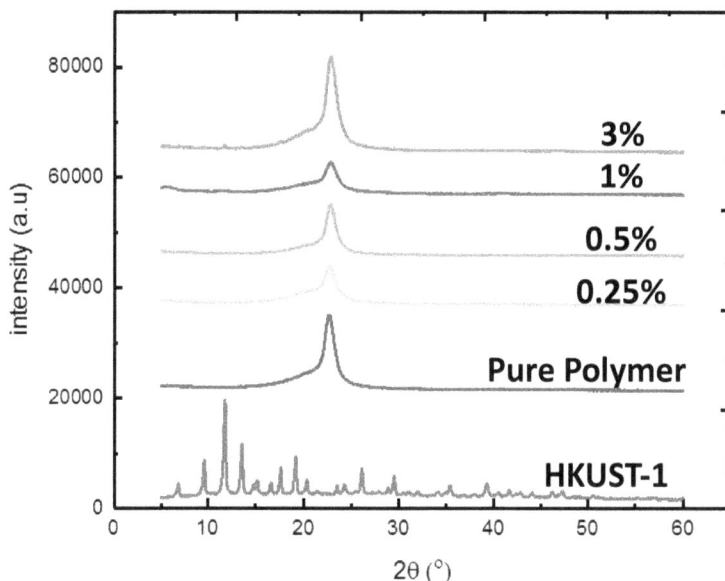

Figure 5.6 WAXD patterns of PEBAX-4533 and MMMs.

The relative crystallinity of different MMMs was also estimated from the ratio of the peak intensities of MMMs and pure polymer, as shown in Table 5.7. From the table, it is clear that the addition of HKUST-1 gradually increased the crystallinity of PEBAX on enhancing the filler fraction, thus, coinciding with DSC findings.

Table 5.7 Relative crystallinities of pure PEBAX-4533 and MMMs

Filler fraction (%)	Relative crystallinity (%)
0	-
0.25	1.25
0.5	1.57
1	1.79
3	2.33

PEBAX comprises of amide and ether segments linked by ester groups, with –OH ends. Among the polymer chains, the amide blocks are linked with strong hydrogen bonds, while the covalent bonds in the polyether repeating units are rather stretchable and bendable. The FTIR spectrum of PEBAX in Figure 5.7 exhibited two characterization peak at 1734 and 1637 cm^{-1} corresponding to C=O stretching vibration of free amide and the hydrogen-bonded amide molecules

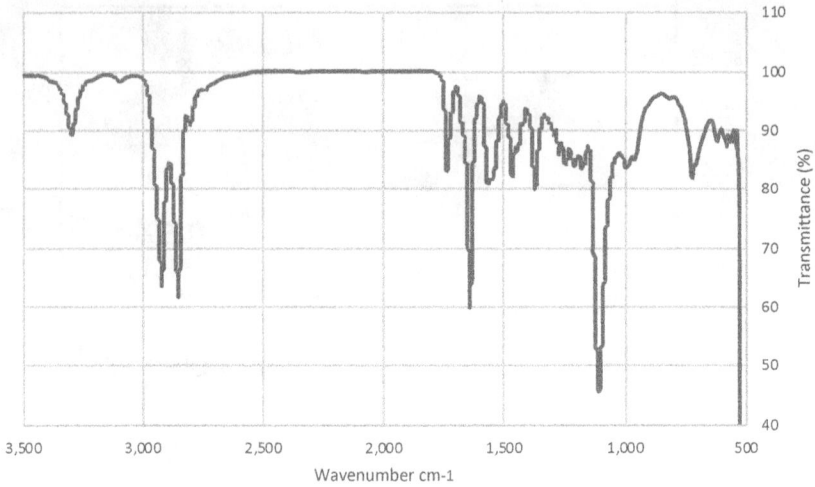

Figure 5.7 FTIR spectrum of PEBAX-4533 film.

respectively. Another peak at 3295 cm^{-1} indicated the presence of N-H group of amide units. For the ether units, a characteristic absorbance band appeared at 1107 cm^{-1} corresponding to the C-O vibration. Moreover, C-H aliphatic groups had a peak at 2918 cm^{-1}. A small bulge appeared around 2796 cm^{-1} which might correspond to the hydrogen bonded -OH groups. The ester units had characteristic peaks at 1734 cm^{-1} for C=O stretching. Figure 5.8 demonstrates the FTIR spectrum

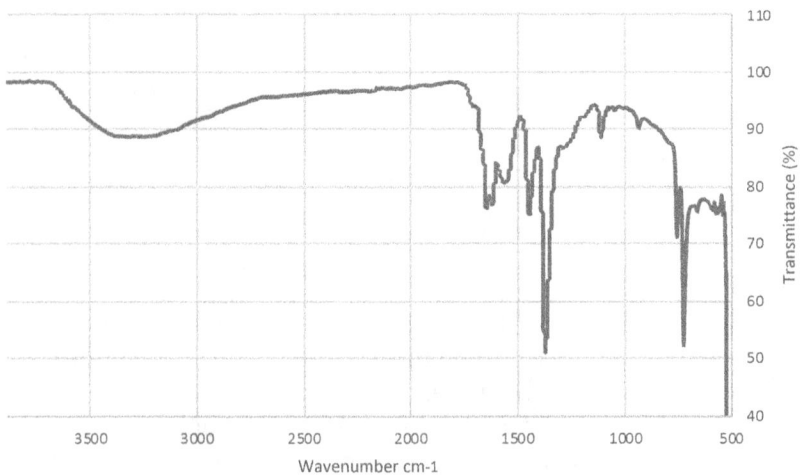

Figure 5.8 FTIR spectrum of HKUST-1 powder.

of HKUST-1. Small peaks around 1668-1643 cm⁻¹ indicated C=O groups of deprotonated benzene tricarboxylic acid. Carboxylate groups have two characteristic peaks owing to two types of stretching: symmetrical with sharp peak at 1370 cm⁻¹ for C=O and anti-symmetrical with wide peak at 1556 cm⁻¹ for O-H. This is due to the aromatic stretching of the para-distributed phenyl groups. The stretching of C-C groups in the ring was observed at 1370 cm⁻¹. Moreover, the presence of a wide band between 1600 and 1500 cm⁻¹ was due to the carboxylate groups forming bidentate inter-cluster bridges that can cause aggregation of the filler particles [15,20]. For the organic ligands, absorbance bands around 1700 cm⁻¹ are related to C=O and C-OH stretching vibration in the free BTC ligands, which were absence in the spectrum. This indicated that the protonated free ligands did not exist in the sample. Several peaks between 1300-500 cm⁻¹ can be attributed to out-of-plane vibrations of the BTC groups such as the stretching of C-O at 111 cm⁻¹ and C-COO at 729 cm⁻¹, 760 cm⁻¹ and 940 cm⁻¹.

Figure 5.9 compares the FTIR spectra of pure PEBAX, HKUST-1 and 3% MMM to analyze the effect of filler on the polymer matrix. It is noticeable that the MMM spectrum overlapped with the spectrum of pure PEBAX. The characteristic peak of HKUST-1 was observed at 1367 cm⁻¹. In general, no new peak was detected in the MMM spectrum, indicating no chemical interactions between the filler and polymer chains. However, the finger-print region in the 500-1000 cm⁻¹

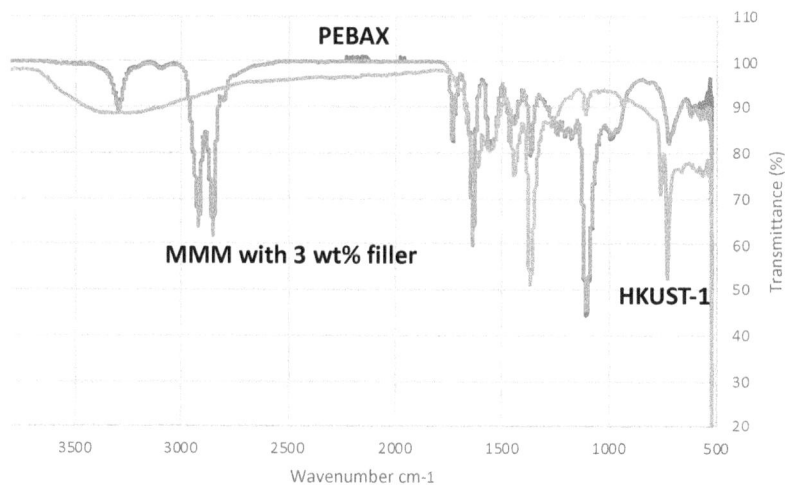

Figure 5.9 FTIR spectra of PEBAX, 3 wt% MMM and HKUST-1.

range indicated the emergence of small peaks, increase in the intensity of the peaks and slight peak shifting, thus, indicating a small degree of physical interactions among the components.

Tables 5.8 and 5.9 summarize the results of the contact angle and surface fee energy analysis (SFE) in DIW and DIM. The results showed

Table 5.8 Images of the contact angle of the droplets on the surface of PEBAX and MMMs with DIM and DIW

MMM filler fraction (%)	DIM	DIW
0		
0.25		
0.5		
3		

Table 5.9 Contact angle and corresponding surface free energy of PEBAX and MMMs

	Contact angle (°)		Surface Free Energy ($\frac{mJ}{m^2}$)		
	DIW	DIM	r	r (d)	r (p)
Pure PEBAX	61.8	32.5	53.5	43.1	10.3
0.25% MMM	64.5	29.3	53.1	44.5	8.6
0.5% MMM	63.6	21.2	55.6	47.4	8.2
3% MMM	64.1	22.7	55.0	47.0	8.1

hydrophilic nature of PEBAX-4533 with relatively low SFE. The MMMs exhibited higher contact angles with water and lower contact angle with DIM, indicating an increase in hydrophobicity due to the non-polar nature of HKUST-1. The MMMs with 0.5 and 3 wt% filler resulted in higher SFE values, while the membrane with 0.25 wt% filler fraction showed a lower SFE. This indicated lower degree of interfacial interactions between the phases in the case of lower filler loadings [21,22]. The increase in SFE of the MMMs is attributed to the higher content of dispersive portions of the filler in the polymer matrix. Oppositely, the polar portions of the SFE were observed to decrease significantly. However, it should be noted that the contact angle analysis should not be taken as a definite measurement due to the limitation associated with polymer morphology. The membrane preparation protocols such as the concentration of the polymer in the solution, type of solvent used and thickness of the film can influence the contact angle values. It should be noted that the highly porous films can result in low contact angle values even though the polymer may be highly hydrophobic.

5.4 Conclusions

Defect-free MMMs were successfully developed using PEBAX-4533 with different weight fractions of HKUST-1 through solvent-evaporation method. Mechanical analysis revealed that the addition of filler decreased both elongation at break and the ultimate tensile strength, however, improvements in Young's modulus were observed with filler addition. The addition of highly crystalline HKUST-1 enhanced the crystallinity of PEBAX and shifted the glass transition temperature to higher values. Furthermore, TGA analysis showed that the MMMs exhibited enhanced thermal stability. FTIR spectra of the MMMs further confirmed the absence of any chemical interactions between the filler and polymer phases. Contact angle measurements showed that the MMMs with high filler content exhibited higher surface free energy.

References

1. Sheth, J. P., Xu, J., and Wilkes, G. L. (2003) Sold state structure property behavior of semicrystalline poly(ether-block-amide) PEBAX (R) thermoplastic elastomers. *Polymer*, **44**(3), 743-756.
2. Solanki S .J., and Desai, R. N. (2013) Polymeric membrane technol-

ogy. *International Journal of Engineering Science and Innovative Technology*, **2**(2), 400-403.

3. Ovid'ko, I. A. (2013) Enhanced mechanical properties of polymer matrix nanocomposites reinforced by graphene inclusions: A review. *Review on Advanced Materials Science*, **34**, 19-25.

4. Venna, S. R., and Carreon, M. A. (2015) Metal organic framework membranes for carbon dioxide separation. *Chemical Engineering Science*, **124**, 3-19.

5. Li, H., Haas-Santo, K., Schygulla, U., and Dittmeyer, R. (2015) Inorganic microporous membranes for H_2 and CO_2 separation - Review of experimental and modeling progress. *Chemical Engineering Science*, **127**, 401-417.

6. Murali, R. S., Sridhar, S., Sankarshana, T., and Ravikumar, Y. V. L. (2010) Gas permeation behavior of PEBAX-1657 nanocomposite membrane incorporated with multiwalled carbon nanotubes. *Industrial & Engineering Chemistry Research*, **49**(194), 6530-6538.

7. Mittal, G., Dhand, V., Rhee, K. Y., Park, S.-J., and Lee, W. R. (2015) A review on carbon nanotubes and graphene as fillers in reinforced polymer nanocomposites. *Journal of Industrial and Engineering Chemistry*, **21**, 11-25.

8. Murali, R. S., Ismail, A. F., Rahman, M. A., and Sridhar, S. (2014) Mixed matrix membranes of Pebax-1657loaded with 4A zeolite for gaseous separations. *Separation and Purification Technology*, **129**, 1-8.

9. Aroon, M. A., Ismail, A. F., Matsuura, T., and Montazer-Rahmati, M. M. (2010) Performance studies of mixed matrix membranes for gas separation: A review. *Separation and Purification Technology*, **75**, 229-242.

10. Basu, S., Cano-Odena, A., and Vankelecom, I. F. J. (2011) MOF-containing mixed-matrix membranes for CO_2/CH_4 and CO_2/N_2 binary gas mixture separations. *Separation and Purification Technology*, **81**(1), 31-40.

11. Lim, S. Y., Choi, J., Kim, H.-Y., Kim, Y., Kim, S.-J., Kang, Y. S., and Won, J. (2014) New CO_2 separation membranes containing gas-selective Cu-MOFs. *Journal of Membrane Science*, **467**, 67-72.

12. Zornoza, B., Tellez, C., Coronas, J., Gascon, J., and Kapteijn, F. (2013) Metal organic framework based mixed matrix membranes: An increasingly important field of research with a large application potential. *Microporous and Mesoporous Materials*, **166**, 67-78.

13. *Thermal Analysis of Polymers - Selected Applications*, Mettler Toledo. Online: https://www.mt.com/dam/Analytical/ThermalAnalysi/TA-PDF/Part%20of%20Polymers-Selected%20Applications.pdf [accessed 16th June 2019].

14. Wee, L. H., Lohe, M. R., Janssens, N., Kaskel, S., and Martens, J. A. (2012) Fine tuning of the metal–organic framework $Cu_3(BTC)_2$

HKUST-1 crystal size in the 100 nm to 5 micron range. *Journal of Materials Chemistry*, **22**, 13742-13746.

15. Sorribas, S., Kudashiva, A., Scholes, C. A., Almedro, E., Zornoza, B., de la Iglesia, O., Tellez, C., and Coronas, J. (2015) Pervaporation and membrane reactor performance of polyimide based mixed matrix membranes containing MOF HKUST-1. *Chemical Engineering Science*, **124**, 37-44.

16. Chui, S. S.-Y., Lo, S. M.-F., Charmant, J. P. H., Orpen, A. G., and Williams, I. D. (1999) A chemically functionalizable nanoporous material $[Cu_3(TMA)_2(H_2O)_3]_n$. *Science*, **283**, 1148-1150.

17. Shahid, S., and Nijmeijer, K. (2014) Performance and plasticization behavior of polymer-MOF membranes for gas separation at elevated pressures. *Journal of Membrane Science*, **470**, 166-177.

18. Sun, B., Kayal, S., and Chakraborty, A. (2014) Study of HKUST (copper benzene-1,3,5-tricarboxylate, Cu-BTC MOF)-1 metal organic frameworks for CH_4 adsorption: An experimental Investigation with GCMC (grand canonical Monte-Carlo) simulation. *Energy*, **76**, 419-427.

19. Chowdhury, P., Bikkina, C., Meister, D., Dreisbach, F., and Gumma, S. (2009) Comparison of adsorption isotherms on Cu-BTC metal organic frameworks synthesized from different routes. *Microporous and Mesoporous Materials*, **117**, 406-413.

20. Li, L., Liu, X. L., Geng, H. Y., Hu, B., Song, G. W., and Xu, Z. S. (2013) A MOF/graphene oxide hybrid (MOF:HKUST-1) material for adsorption of methylene blue for aqueous solution. *Journal of Materials Chemistry A*, **1**, 10292-10299.

21. Selvi, M., Vengatesan, M. R., Devaraju, S., Kumar, M., and Alagar, M. (2014) In situ sol–gel synthesis of silica reinforced polybenzoxazine hybrid materials with low surface free energy. *RSC Advances*, **4**, 8446-8452.

22. Jainczuk, B., and Bialopiotrowicz, T. (1990) The total surface free energy and the contact angle in the case of low energetic solids. *Journal of Colloid and Interface Science*, **140**, 362-372.

6

Bio-Polyamide/Doped Graphene Nanocomposites

6.1 Introduction

Among engineering plastics, polyamides (PAs) have dominant applications in various fields, such as construction, electronics, automotive, consumer goods and sports equipment. PAs exhibit extraordi-
nary characteristics such as good abrasion resistance, processability
and mechanical performance. Polyamide-6 and polyamide-66 are the
prominent PAs, mainly employed in the preparation of fibers [1-4].

Development of polymer nanocomposites with a variety of nanofillers (like graphene, carbon nanotubes or cellulose nanocrystals) is
generally employed to further enhance the polymer properties [5-8].
Graphene has been recognized as one of the most promising carbon
materials in the recent years [9-13]. N-doped graphene is a new kind
of hybrid material with efficient structural properties, including high
surface area, accessibility and conductivity, thus, immensely use for
electrochemical applications [14]. Doping with nitrogen atoms has
drawn a lot of interest compared to the other elements in order to
enhance the electrochemical behavior of catalysts based on graphene
[15]. It has also been shown in recent studies that p-doped graphene
has a better catalytic behavior as compared to the n-doped graphene
owing to its higher ability for electron donation and large covalent
radius [16]. Thus, p-doped graphene has also been studied for use in
energy storage/generation fields comprising of fuel cells [17], lithium
ions batteries [18] and supercapacitors [19].

In some studies utilizing polyamides, Baek *et al.* [20] prepared polyamide nanocomposites incorporated with carbon nanotubes and
used these as reverse osmosis membranes. The membranes displayed almost 30% enhancement in water flux. Zhu *et al.* [21] reported bio-based PA nanocomposites incorporated with cellulose
nanofibers with improved thermal and mechanical properties. In a
recent study, Mirzakhanjan *et al.* [22] also synthesized bio-based PA

Swati Singh and Vikas Mittal, The Petroleum Institute (part of Khalifa
University of Science and Technology), Abu Dhabi, UAE
Current address: Bletchington, Wellington County, Australia

nanocomposites with acid-functionalized carbon nanotubes. Meng *et al.* [23] prepared the composites containing graphene modified by caprolactam onium ion and PA6 matrix through *in-situ* intercalation polymerization, and the composites exhibited high thermal conductivity. Ren *et al.* [24] reported that the fabric composites with PS-graphene filler exhibited enhanced tribological properties. The tribological properties were observed to be dependent on the fillers and test conditions [25].

In this study, the preparation and characterization of bio-nanocomposites, obtained through solution blending of bio-polyamide and doped graphene (n and p), have been presented. For comparison, the composites were also prepared with pristine graphene.

6.2 Experimental

6.2.1 Materials

Renewably sourced biodegradable polyamide was provided by DuPont. Graphene oxide (GO) was synthesized in the laboratory by following modified Tour's method. The thermally reduced graphene (TRG) nanoplatelets with the product name N002-PDR were purchased from Angstron Materials, USA. The material possessed a density of ≤ 2.20 g/cm^3 and specific surface area in the range 400-800 m^2/g. Urea, triphenyl phosphate, dimethyl formamide (DMF) and ethanol were procured from Sigma Aldrich.

6.2.2 Synthesis of n-doped Graphene (NG)

500 mg GO was dispersed in 500 mL water using ultrasonication. Subsequently, 6 g urea was added to the dispersion. The contents were kept continuously stirred at room temperature for 24 h. Afterwards, the solution was vacuum dried, and the dried powder was exfoliated at 600 °C for 5 min.

6.2.3 Synthesis of p-doped Graphene (PG)

500 mg GO was added to a mixture of 125 mL ethanol and 125 mL DMF. The mixture was sonicated until a clear dispersion was obtained. 2 g triphenyl phosphate was subsequently added to the mixture. The solution was then kept under vigorous stirring for 2 h at 100 °C. Afterwards, the mixture was left to cool down and was washed

continuously with water and ethanol. The precipitate was subsequently dried in a vacuum oven for 12 h at 70 °C. Finally, the dried powder was exfoliated at 600 °C for 5 min.

6.2.4 Synthesis of Nanocomposites using Solution Blending

DMF was used as a solvent for the solution blending process. For the preparation of a batch of 8 g of the nanocomposite containing 1 wt% n-doped graphene, 0.08 g of n-doped graphene was added to 300 mL DMF. The dispersion was kept under ultrasonication for 3 d. Afterwards, 7.92 g of bio-PA was added to the dispersion, followed by stirring at 300 °C for 2 h. The solution was then poured into a container and left for drying. Similarly, other nanocomposites with different concentrations of n-doped graphene, p-doped graphene and TRG were prepared. For comparison, bio-PA was also processed similarly. The samples were named as BS1 for pure bio-PA; BS2, BS3 and BS4 for bio-PA incorporated with 1 wt%, 2wt% and 5wt% TRG respectively; BS5, BS6 and BS7 for corresponding composites with p-doped graphene and BS8, BS9 and BS10 for corresponding composites with n-doped graphene. The test specimens for tensile analysis were prepared by using injection molding (Thermo Scientific Haake Minijet Pro). The mold and cylinder temperature were fixed as 125 °C and 235 °C, respectively. Further, the injection and post pressure were set as 570 and 470 bars, respectively, for 10 s.

6.2.5 Characterization

Universal testing machine (Instron 5567) was used for the tensile analysis. The dumbbell shape samples of 4 mm width, 53 mm length and 2 mm thickness were used for the measurements at room temperature with a loading rate of 10 mm/min using Win Test Analysis software. An average of five specimens was reported.

Calorimetric properties were studied using dynamic scanning calorimetry (Discovery TA DSC) in nitrogen atmosphere. 3-7 mg of the samples was used for the analysis. The samples were heated from 35 to 200 °C and cooled from 200 to -60 °C, using a rate of 10 °C/min. The second heating and cooling scans were similarly performed and used for obtaining the calorimetric properties.

Wide angle X-ray diffraction (WAXRD) patterns of the fillers and polymeric materials were measured on a Panalytical powder diffractometer in reflection mode with CuKα radiation (λ = 1.5406 Å). The

samples were scanned at room temperature from 2θ = 5-60° with a
step size of 0.02° and step time of 10 s. Zero-background holders
were used to minimize the noise.

Raman analysis of the fillers and nanocomposite samples was carried out using a confocal microRaman spectrometer (Horiba LabRAM
HR), equipped with a 633 nm laser source.

6.3 Result and Discussion

X-ray diffractograms of the fillers are presented in Figure 6.1. The
changes induced by the fillers on the crystal structure of PA were

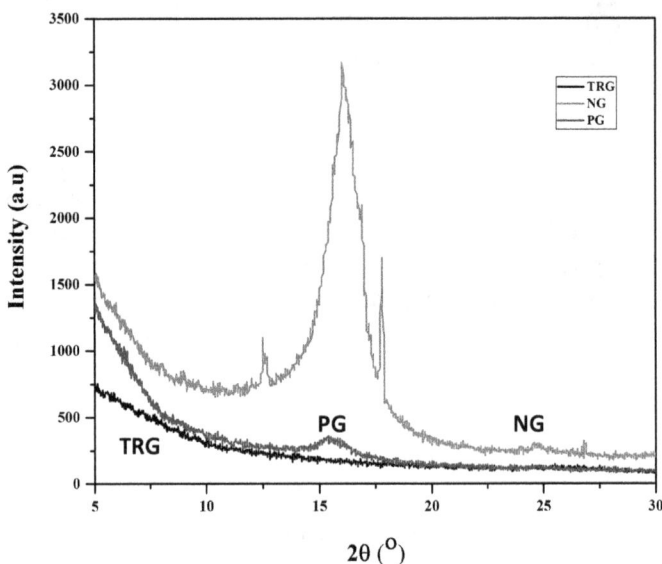

Figure 6.1 XRD spectra of TRG, NG and PG.

analyzed further from the XRD patterns of PA and PA nanocomposites. Two characteristic peaks at 2θ = 20.5° and 23.7° [26-29] were
observed for all polymeric materials. The filler characteristic peaks
were also absent in the diffractograms of the composites, thus, indicating a good degree of filler dispersion. In addition, no new peaks
were observed, thus, implying no significant interaction between the
PA and filler phases.

Calorimetric properties of pure PA and nanocomposites (peak
melting temperature (H_m), peak crystallization temperature (H_c),
melting enthalpy (ΔH_m) and crystallization enthalpy (ΔH_c)) are listed

in Table 6.1. Two different melting transitions were observed in the thermograms and were noted as m1 and m2. This might be due to the co-existence of the alpha and gamma crystal structures of PA or due to the melting-recrystallization procedure during the second heating scan [26-32]. Nonetheless, the peak for T_m at 190 °C reduced as the concentration of graphene increased. The result probably indicated that the crystal structure slowly changed weakly from alpha to a typical gamma structure [33-37] or the melting-recrystallization step moderately disappeared. No noteworthy change in the peak crystallization temperature was observed on incorporating the nanofillers.

Table 6.1 Calorimetric properties of the pure PA and PA nanocomposites

	T_{m1} (°C)	T_{m2} (°C)	T_c (°C)	ΔH_{m1}	ΔH_{m2}	ΔH_c
BS1	190	200	178	2.2	4.1	10.3
BS2	190	200	179	0.6	5.6	9.7
BS3	190	200	180	0.3	6.1	8.6
BS4	190	201	182	0.2	7.6	11.0
BS5	199	201	179	2.9	5.7	10.1
BS6	190	200	178	2.1	4.8	9.3
BS7	190	200	178	2.2	4.9	9.0
BS8	190	200	178	2.5	5.7	10.7
BS9	190	199	178	3.0	5.0	10.1
BS10	190	201	177	1.0	4.4	11.9

Mechanical properties of the materials are outlined in Table 5.2. Pure polyamide had a tensile modulus of 1192 MPa, which was enhanced to 1915 MPa for 2% TRG composite, thus, exhibiting an increase of 60%. The enhancements could be attributed to the homogeneous filler dispersion, strong interfacial adhesion and effective load transfer to thermally reduced graphene. However, the value decreased to 1458 MPa for 5% TRG, which might have been caused by the filler agglomeration. In the case of nanocomposites incorporated with p-doped graphene, a gradual decrease in the modulus was observed on enhancing the filler fraction. The composites with n-doped graphene (NG) exhibited the highest modulus value at 2% filler content, followed by a decrease in the modulus at 5% filler content.

Table 5.2 Mechanical properties of pure PA and PA nanocomposites

	UTS (MPa)	Modulus (MPa)	Extension at Break (%)
BS1	50	1192	150
BS2	49	1507	8
BS3	44	1915	6
BS4	31	1458	2
BS5	50	1346	20
BS6	42	1193	17
BS7	41	956	11
BS8	51	1269	22
BS9	50	1335	19
BS10	51	1273	17

A significant decrease in tensile strain was noted for all nanocomposites as compared to pristine polymer. The value decreased from 150% to 8% for BS2, which was decreased further on enhancing the amount of filler. In the case of NG and PG based nanocomposites, though the elongation decreased as a function of filler content, however, it was still appreciable. The tensile strength of the TRG and PG composites also exhibited a minor reduction with filler content.

Raman spectra of the fillers is presented in Figure 6.2. Raman spectroscopy is extensively used to study the disorder and defects in graphene samples [38]. The Raman spectrum of TRG exhibited a D band at 1339 cm^{-1}, corresponding to sp^3- hybridized carbons and assigned to the breathing mode of κ-point phonons of A_1g symmetry. The G band at 1595 cm^{-1} corresponded to sp^2- hybridized carbons and allocated to the first-order scattering of E_2g phonons [39-41]. Small changes in the relative ratio of these bands were observed in the case of NG and PG, which would have contributed to the observed behavior of these materials in the nanocomposites.

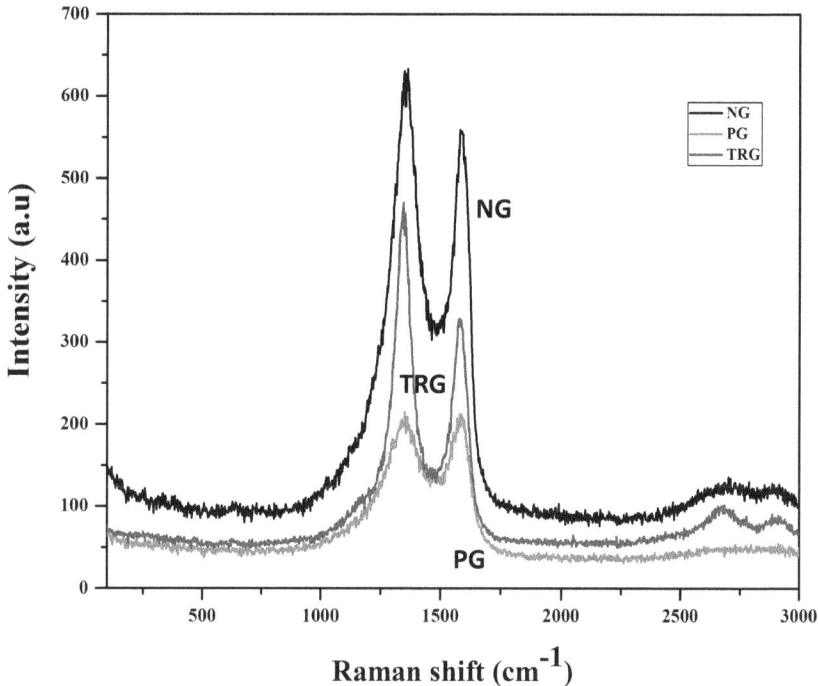

Figure 6.2 Raman spectra of TRG, NG and PG.

6.4 Conclusion

In this study, three different nanofillers were employed to synthesize bio-polyamide based nanocomposites via simple solution mixing. The tensile modulus of the nanocomposites exhibited a significant improvement, especially at lower filler fractions. Highest improvement in the modulus was observed for the composites with thermally reduced graphene (TRG), however, these composites suffered from significant brittleness. N-doped and p-doped graphene based composites though did not exhibit the similar extent of modulus enhancement as TRG, however, their elongation was still appreciable, and the loss in tensile strength was minimal. Calorimetric analysis also exhibited two different melting transitions in the nanocomposites. Addition of filler was observed to have no change on the crystallization temperature of the nanocomposites. Also, the filler peaks were absent in the diffraction patterns of the nanocomposites, thus, indicating a good degree of filler dispersion. The future studies on these composites involve in-depth analysis of conductivity enhancement owing to doping.

References

1. Feng, C., Khulbe, K. C., Matsuura, T., Tabe, S., Ismail, A. F., (2013) Preparation and characterization of electro-spun nanofiber membranes and their possible applications in water treatment. *Separation and Purification Technology*, **102**, 118-135.
2. Rahimi, S. K., and Otaigbe, J. U. (2016) The role of particle surface functionality and microstructure development in isothermal and non-isothermal crystallization behavior of polyamide 6/cellulose nanocrystals nanocomposites. *Polymer*, **107**, 316-331.
3. Aussawasathien, D. Teerawattananon, C., and Vongachariya, A. (2008) Separation of micron to sub-micron particles from water: Electrospun nylon-6 nanofibrous membranes as pre-filters. *Journal of Membrane Science*, **315**(1-2), 11-19.
4. Tan, K., and Obendorf, S. K. (2007) Fabrication and evaluation of electrospun nanofibrous antimicrobial nylon 6 membranes. *Journal of Membrane Science*, **305**(1-2), 287-298.
5. Kim, H. S., Jin, H.-J., Myung, S. J., Kang, M., Chin, I.-J. (2006) Carbon nanotube-adsorbed electrospun nanofibrous membranes of nylon 6. *Macromolecular Rapid Communications*, **27**(2), 146-151.
6. Pant, H. R., Bajgai, M. P., Nam, K. T., Seo, Y. A., Pandeya, D. R., Hong, S. T., and Kim, H. Y. (2011) Electrospun nylon-6 spider-net like nanofiber mat containing TiO2 nanoparticles: A multifunctional nanocomposite textile material. *Journal of Hazardous Materials*, **185**(1), 124-130.
7. Pant, H. R., Pandeya, D. R., Nam, K. T., Baek, W.-I., Hong, S. T., and Kim, H. Y. (2011) Photocatalytic and antibacterial properties of a TiO_2/nylon-6 electrospun nanocomposite mat containing silver nanoparticles. *Journal of Hazardous Materials*, **189**(1-2), 465-471.
8. Pant, H. R., Park, C. H., Tijing, L. D., Amarjargal, A., Lee, D.-H., and Kim, C. S. (2012) Bimodal fiber diameter distributed graphene oxide/nylon-6 composite nanofibrous mats via electrospinning. *Colloids and Surfaces A: Physicochemical and Engineering Aspects*, **407**, 121-125.
9. Rogers, J. A., Someya, T., and Huang, Y. (2010) Materials and mechanics for stretchable electronics. *Science*, **327**(5973), 1603-1607.
10. Zhu, Y., Murali, S., Cai, W., Li, X., Suk, J. W., and Potts, J. R. (2010) Graphene and graphene oxide: synthesis, properties, and applications. *Advanced Materials*, **22**(35), 3906-3924.
11. Stankovich, S., Dikin, D. A., Dommett, G. H., Kohlhaas, K. M., Zimney, E. J., and Stach, E. A. (2006) Graphene-based composite materials. *Nature*, **442**(7100), 282-286.

12. Lee, C., Wei, X., Kysar, J. W., Hone, J. (2008) Measurement of the elastic properties and intrinsic strength of monolayer graphene. *Science*, **321**(5887), 385-388.

13. Berman, D., Erdemir, A., Sumant. A. V. (2014) Graphene: A new emerging lubricant. *Materials Today*, **17**(1), 31-42.

14. Liang, J., Du, X., Gibson, C., Du, X. W., and Qiao, S. Z. (2013) N-doped graphene natively grown on hierarchical ordered porous carbon for enhanced oxygen reduction. *Advanced Materials*, **25**, 6226-6231.

15. Wang, X. W., Sun, G. Z., Routh, P., Kim, D. H., Huang, W., and Chen, P. (2014) Heteroatom-doped graphene materials: syntheses, properties and applications. *Chemical Society Reviews*, **43**, 7067-7098.

16. Ngidi, N. P. D, Ollengo, M. A., and Nyamori, V. O. (2018) Heteroatom-doped graphene and its application as a counter electrode in dye-sensitized solar cells. *International Journal of Energy Research*, **43**, 1702-1734.

17. Zhang, C., Mahmood, N., Yin, H., Liu, F., and Hou, Y. (2013) Synthesis of phosphorus-doped graphene and its multifunctional applications for oxygen reduction reaction and lithium ion batteries. *Advanced Materials*, **25**, 4932-4937.

18. Ma, X., Ning, G., Qi, C., Xu, C., and Gao, J. (2014) Phosphorus and nitrogen dual-doped few-layered porous graphene: a high-performance anode material for lithium-ion batteries. *ACS Applied Materials and Interfaces*, **6**, 14415-14422.

19. Wen Y., Wang, B., Huang, C., Wang, L., and Hulicova-Jurcakova D. (2015) Synthesis of phosphorus-doped graphene and its wide potential window in aqueous supercapacitors. *Chemistry A European Journal*, **21**, 80-85.

20. Baek, Y., Kim, H. J., Kim, S. H., Lee, J. C., and Yoon, J. (2017) Evaluation of carbon nanotube-polyamide thin-film nanocomposite reverse osmosis membrane: Surface properties, performance characteristics and fouling behavior. *Journal of Industrial and Engineering*, **56**, 327-334.

21. Zhu, J. H., Kiziltas, A., Lee, E. C., and Mielewski, D. (2015) Bio-based Polyamides Reinforced with Cellulose Nanofibers - Processing and Characterization. *SPE ANTEC*, USA.

22. Mirzakhanian, Z., Hasani, M., and Faghihi, K. (2018) Synthesis of bio-based polyamide/acid-functionalized multiwalled carbon nanotube nanocomposites using vanillin. *Polymer-Plastics Technology and Engineering*, **57**(13), 1367-1376.

23. Meng, F., Huang, F., Guo, Y., Chen, J., Chen, X., Hui, D., He, P., Zhou, X., and Zhou, Z. (2017) In situ intercalation polymerization approach to polyamide-6/graphite nanoflakes for enhanced thermal conductivity. *Composites Part B: Engineering*, **117**, 165-173.

24. Ren, G., Zhang, Z., Zhu, X., Ge, B., Guo, F., and Men, X. (2013) Influence of functional graphene as filler on the tribological behaviors of No-

mex fabric/phenolic composite. *Composite Part A*, **49**, 157-164.

25. Bijwe, J., Naidu, V., Bhatnagar, N., and Fahim, M. (2006) Optimum concentration of rein- forcement and solid lubricant in polyamide 12 composites for best triboperformance in two wear modes. *Tribology Letters*, **21**(1), 57-64.

26. Li, Y., Yan, D., and Zhu, X. (2000) Crystalline transition in Nylon 10 10, *Macromolecular Rapid Communications*. **21**, 1282-1285.

27. Xu, Z., and Gao, C. (2010) In situ polymerization approach to graphene-reinforced nylon-6 composites, *Macromolecules*, **43**, 6716-6723.

28. Li, Y., and Yan, D. (2001) Preparation, characterization and crystalline transition behaviors of polyamide 4 14, *Polymer*, **2**, 5055-5058.

29. Yazdani, B., Xia, Y., Ahmad, I., and Zhu Y. (2015) Graphene and carbon nanotube (GNT)-reinforced alumina nanocomposites. *Journal of the European Ceramic Society*, **35**, 179-186.

30. Zhang, X., Fan, X., Li, H., and Yan C. (2012) Facile preparation route for graphene oxide reinforced polyamide 6 composites via in situ anionic ring-opening polymerization. *Journal of Materials Chemistry*, **22**(45), 24081-24091.

31. Bhattacharyya, A. R., Pötschke, P., Häußler, L., Fischer, D. (2005) Reactive compatibilization of melt mixed PA6/SWNT composites: mechanical properties and morphology. *Macromolecular Chemistry and Physics*, **206**(20), 2084-2095.

32. Penel-Pierron, L., Depecker, C., Seguela, R., and Lefebvre, J. M. (2001) Structural and mechanical behavior of nylon 6 films part I. Identification and stability of the crystalline phases. *Journal of Polymer Science, Part B: Polymer Physics*, **39**(5), 484-495.

33. Xu, Z., and Gao, C. (2010) In situ polymerization approach to graphene-reinforced nylon-6 composites. *Macromolecules*, **43**(16), 6716-6723.

34. Arimoto, H. (1964) Transition of nylon 6. *Journal of Polymer Science, Part A: General Papers*, **2**(5), 2283-2295.

35. Medellin-Rodriguez, F., Larios-Lopez, L., Zapata-Espinoza, A., Davalos-Montoya, O., Phillips, P., and Lin, J. (2004) Melting behavior of polymorphics: molecular weight dependence and steplike mechanisms in nylon-6. *Macromolecules*, **37**(5), 1799-1809.

36. Song, N., Yang, J., Ding, P., Tang, S., and Shi, L. (2015) Effect of polymer modifier chain length on thermal conductive property of polyamide 6/graphene nanocomposites. *Composites Part A*, **73**, 232–241.

37. Chen, M., Yin, J., Jin, R., Yao, L., Su, B., and Lei, Q. (2015) Dielectric and mechanical properties and thermal stability of polyimide graphene oxide composite films. *Thin Solid Films*, **584**, 232-237.

38. Tan, P. H., Han, W. P., Zhao, W. J., Wu, Z. H., Chang, K., Wang, H., Wang, Y. F., Bonini, N., Marzari, N., Pugno, N., Savini, G., Lombardo, A., and

Ferrari, A. C. (2012) The shear mode of multilayer graphene. *Nature Materials*, **11**(4), 294-300.

39. Tang, Z., Kang, H., Shen, Z., Guo, B., Zhang, L., and Jia, D. (2012) Grafting of polyester onto graphene for electrically and thermally conductive composites. *Macromolecules*, **45**(8), 3444-3451.

40. He, D., Jiang, Y., Lv, H., Pan, M., and Mu, S. (2013) Nitrogen-doped reduced graphene oxide supports for noble metal catalysts with greatly enhanced activity and stability. *Applied Catalysis B: Environmental,* **132**, 379-388

41. Cao, L., Sun, Q., Wang, H., Zhang, X., and Shi H. (2015) Enhanced stress transfer and thermal properties of polyimide composites with covalent functionalized reduced graphene oxide. *Composites Part A*, **68**, 140-148.

7

Anti-microbial Polymeric Materials and Coatings

7.1 Introduction

Controlling the harmful effects of micro-organisms is an important task for improving human health. Major health problems can occur due to the uncontrolled growth of microbes. [1-3]. Most of the polymers and conventional materials used in day to day life generally exhibit no response towards the growth of micro-organisms. Some of them can even act as a medium for the accumulation and propagation of microbes. The surrounding environment also helps in the proliferations and multiplication of these micro-organisms. In this respect, development of anti-microbial materials is very critical [4-6]. Anti-microbial agents can be broadly defined as materials that are capable of neutralizing micro-organisms and preventing their further growth. Low molecular anti-microbial agents are widely used in sterilization of water, anti-microbial drugs, food preservation and soil sterilization [7-16]. The polymeric materials with anti-microbial properties are generally termed as polymeric biocides. The polymeric biocides can be present in many forms like polymeric biocide incorporated fibers or directly extruded as fibers. These fibers can be used in sterile bandages and clothing materials [17-20].

Another approach to inhibit the growth of micro-organisms is the development of anti-microbial surfaces and coatings. These coatings have two-fold action which includes the repulsion of microbes on the surface and neutralizing the microbes in contact with the surface [21-24]. The action is generally achieved by using biocides such as triclosan active chlorine, antibiotics, anti-microbial ammonium, silver compounds, etc. These materials work efficiently in eliminating the microbes, however, can exhaust after continuous usage. This can be prevented by creating surfaces that can produce the biocides catalytically. The catalytic activity can be induced through a number of external stimuli like heat, electricity, optical energy, etc. Another way of

Liyamol Jacob and Vikas Mittal*, The Petroleum Institute (part of Khalifa University of Science and Technology), Abu Dhabi, UAE
*Current address: Bletchington, Wellington County, Australia

preventing the exhaustion of biocides is the use of materials that become active only during a microbial contact. These are called contact-active anti-microbial surfaces, and are usually developed using anti-microbial polymers tethered onto the surface [25-28].

7.2 Different Anti-microbial Polymeric Materials and Coatings

Timofeeva and Kleshcheva [29] recently reviewed the trends in polymeric biocides. Polymeric biocides have been synthesized as a large class of copolymers, either quaternized or functionalized with bioactive groups, anti-microbial macromolecular systems and inherent biocidal polymers [30-36]. In one such study, cycloamine monomer, 3-allyl-5,5-dimethylhydantoin (ADMH), was grafted onto high performing fiber surfaces using a continuous "pad-dry-cure" technique [37]. ADMH readily grafted onto the fibers in the presence of polymers like poly(ethylene glycol)-diacrylate (PEG-DIA). The hydantoin structures of the grafted polymer exhibited regeneratable and long lasting anti-bacterial properties on exposure to chlorine. The polymer was effective against both gram positive and gram negative bacteria. Nigamatullin *et al.* [38] fabricated novel polymer biocides using clay and polymer nanotechnology. The anti-bacterial activity of the cationic surfactant modified organoclay was studied for Staphylococcus aureus (gram positive) and Escherichia coli (gram negative) bacteria. The organoclay's interaction with the cell surface was confirmed to be the reason behind the anti-microbial activity.

In another study, amination was used to introduce functional groups on polyacrylonitrile (PAN) [39]. The aminated polymer was further treated with benzaldehyde and its derivatives to evolve anti-microbial properties. The benzaldehyde derivatives used for immobilization on the aminated polymer included 2,4-dihydroxybenzaldehyde and 4-hydroxybenzaldehyde. Staphylococcus aureus, Pseudomonas aeruginosa, Escherichia coli, Salmonella typhi, Cryptococcus neoformans, Candida albicans, Aspergillus niger and Aspergillus flavus were used to study the anti-microbial activity using viable cell counting and cut plug method. It was concluded that the aminated polymer had a durable efficiency in eliminating bacteria and fungi. An increase in the number of phenolic hydroxyl groups of the bioactive moiety was also observed to enhance the biocidal efficiency of the material.

Fluorescence depolarization method was used to study the interaction of poly(hexamethylene biguanide hydrochloride) (PHMB) with phospholipid bilayers [40]. PHMB is a polymer biocide containing biguanide groups in its main chain. The neutral phosphatidylcholine (PC) remained unaffected by the interaction, while the negatively charged bilayers, made by phosphatidylglycerol (PG) alone or mixture of PC of PG, were detected. In the gel phase, the addition of PHMB reduced the fluorescence polarization of diphenylhexatriene embedded in the negatively charged bilayers to a large extent. Polytriazoles (PTAs) based on soyabean oil were also prepared using click polymerization method by Gholami *et al.* [41]. Two different propyn-functional urethane monomers were reacted with azidated soybean oil (ASBO). To control the biological activity, Cloisite 30B (MMT-N3) and azidated derivatives of a quaternary pyridinium salt (APS) were also grafted on PTAs. PTAs showed a good cytocompatibility towards fibroblast cells in the range of 86 to 98 percentage viability. C. albicans, P. aeruginosa and S. aureus were used to study the anti-bacterial and anti-fungal properties of PTAs. The mutual usage of the modifier materials in coatings showed a high biocidal activity of up to 99% reduction.

Photodynamic therapy (PDT) was used to estimate the anti-bacterial activity of novel oligo(thiophene ethynylene) (OTE) against Escherichia coli, Ralstonia solanacearum, Staphylococcus epidermidis and Staphylococcus aureus in vitro [42]. OTE showed remarkably broad spectrum of high anti-bacterial activity after white light irradiation. 52, 24, 13 and 8 ng/mL half inhibitory concentration (IC50) values were obtained for R. solanacearum, E. coli, S. epidermidis and S. aureus respectively after 30 min white light irradiation. The mechanism of action of this material is depicted in Figure 7.1. Particularly, at a concentration of 180 ng/mL, OTE exhibited a strong and specific dark killing capability against S. aureus for 30 min.

In another report among several other novel studies on polymeric biocides [43-50], a bacterial cellulose based material was combined with the polymeric biocide polyhexamethylene guanidine hydrochloride (PHMG-Cl) into a porous structure [50]. The bacterial cellulose exhibited effective saturation in the polymer scaffold. The modified PHMG-Cl biocide bacterial cellulose film specimens exhibited effective efficiency against a range of micro-organisms like yeast Pseudomonas syringae PV. Tomato DC 3000, phytopathogenic Xanthomonas campestris PV. Campestris IMBG 299, Klebsiella pneumonia IMBG 233 and Staphylococcus aureus. Further, the material prevented biofilm formation by prokaryotic and eukaryotic organisms. The biocidal activity of the composite depended on the rate of biocide release.

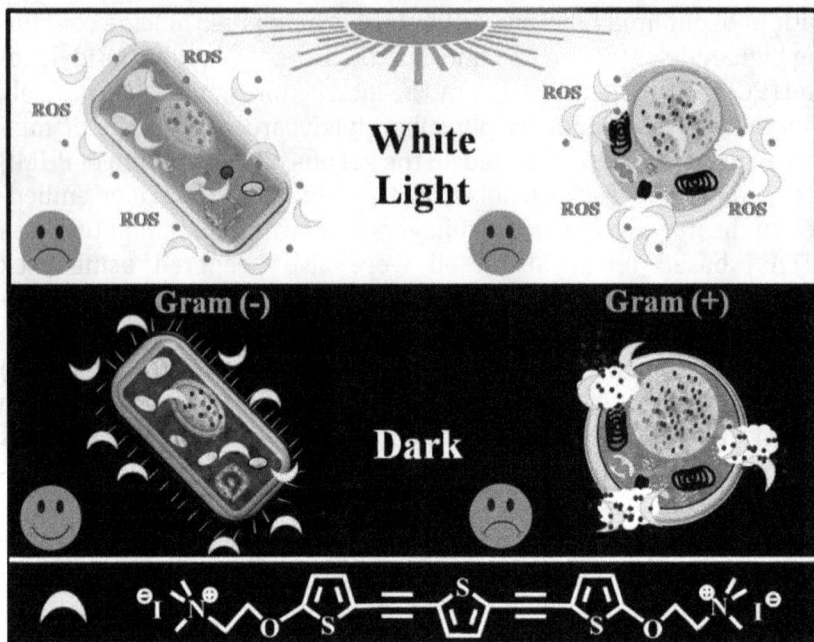

Figure 7.1 Mechanism for OTE cytotoxicity under white light and in dark. Reproduced from Reference 42 with permission from American Chemical Society.

Mycobacteria is one of the most prominent micro-organisms causing a variety of diseases and even mortality. It has resistance towards most of the commonly used disinfectants including low molecular quaternary polymeric biocides owing to its distinct cell wall structure. For studying the biocidal activity on mycobacteria, non-quaternary protonated polydiallylamines (PDAAs) were synthesized [51]. The non-quaternary PDAA acted by breaking the inner membrane permeability in M. Smegmatis cells within 20 min of contact. In another study, bromoacetyl chloride or chloroacetyl chloride modified chitosan were synthesized [52]. Triphenyl phosphine and triethyl amine were used to quaternize the haloacetylated chitosan derivatives. The new derivatives exhibited much better efficiency than chitosan and other commonly used derivatives against gram positive and gram negative bacteria and fungi. The death of the cells was achieved by the complete leakage of cytoplasm. Charged metallopolymers have also been studied for the effective lysing of bacterial cell walls and inhibiting the activity of β-lactamase in Methicillin-resistant Staphylococcus aureus (Figure 7.2) [53]. The formation of unique ion pairs between the cationic cobaltocenium moieties and carboxylate anions inhibited β-lactamase hydrolysis in various β-lactam antibiotics, including cefazolin, amoxicillin ampicillin and penicillin-G. Other biocides with an ability to neutralize multidrug resistant bacteria have also been reported [54-57].

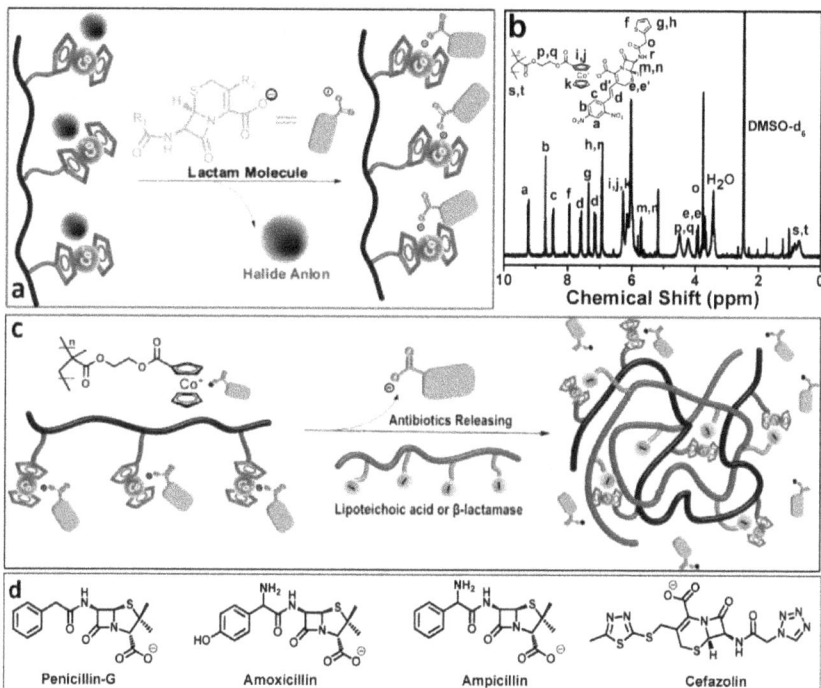

Figure 7.2 (a) Cationic cobaltocenium-containing polymers and β-lactam forming ion-pairs between antibiotics, (b) cationic cobaltocenium-containing polymer and nitrocefin ion-pair's 1 H NMR spectrum, (c) β-lactamases and lipoteichoic acid releasing antibiotic from antibiotic-metallopolymer ion pairs and (d) β-lactam antibiotics utilized in the study. Reproduced from Reference 53 with permission from American Chemical Society.

The formation of biofilms make the destruction of micro-organisms almost 1000 times more difficult. Out of the various methods to prevent the formation of biofilms, the use of coatings capable of releasing anti-microbial agents and containing anti-adhesive components is common. Toxicity of the released anti-microbial agent towards microbes is important in healthcare. Anti-adhesive coatings, on the hand, generally do not interfere with the growth, instead these make it difficult for the biofilm to form. In the body, protein adhesion is the predecessor for any bodily bacterial adhesion, and anti-adhesive coatings are designed to be resistant to such protein adhesion [58-62]. In another study combining the use of two or more operational entities, Gehring *et al.* [63] reported mesoporous organosilicas (PMOs) based on co-condensation of sol-gel precursors with bridging phenyl derivatives $R_{F1,2}C_6H_3[Si(O^{iso}Pr)_3]_2$ (Figure 7.3). Subsequently, PMOs containing high density of thiol and sulfonic acid units were prepared as mesoporous nanoparticles. Another study has also reported the use of triazole/nanotubes conjugates as filler for gel-coat nanocomposites [64].

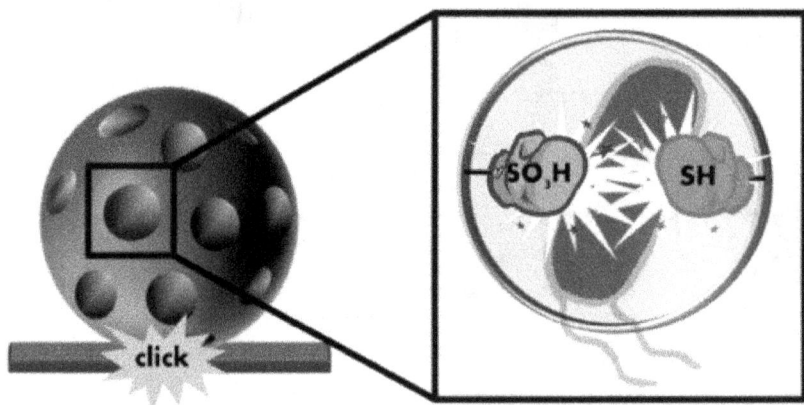

Figure 7.3 Schematic representation of the superacid and click functionalities. Reproduced from Reference 63 with permission from American Chemical Society.

Finlay *et al.* [65] reported biofilms of Navicula incerta prepared from a novel culture system using an open channel flow with adjustable bed shear stress values (0-2.4 Pa). Polydimethylsiloxane elastomer (PDMSe) and glass were used to study and differentiate the biofilm development. The growth of biofilm was prevented by a critical shear stress of 1.3-1.4 Pa on glass. In the case of elastomer, even at 2.4 Pa, the biofilm continued to grow. Schultz *et al.* [66] also presented skin-friction results for fouling-release (FR) hull coatings and suggested a new effective roughness length scale (keff) for biofilms. Ciliate assemblages have also been studied to compare anti-fouling (AF) and FR coatings [67]. In microbial fouling, toxicity of AF coatings is an issue as compared to the FR coatings. The toxic effects shown by AF coatings caused important differences in ciliate species assembly.

In another study, biocidal coatings like controlled depletion polymer (CDP), self-polishing hybrid (SPH) and self-polishing copolymer (SPC) were studied for the microbial community composition of the biofilms developed on them after immersion in sea for a year [68]. Distinct bacterial structures on various coatings were revealed by pyrosequencing of 16S rRNA's genes.

Polymer biocides have considerable importance in textile industry [69]. The need of enhanced hygiene products has paved the way for anti-microbial textiles which have found their use in protective clothing, home furnishing, sportswear and wound-dressings [70-76]. Simultaneous one-step sono-chemical deposition of chitosan and ZnO nanoparticles (NPs) on cotton fabric to achieve anti-microbial textiles was reported by Petkova *et al.* [77]. The biocompatibility and anti-microbial efficacy were further improved by optimizing the reaction duration and concentration. 2 mM ZnO NPs suspension in a 30 min sono-chemical process imparted the cotton fabric the highest anti-bacterial activity. This ability was further enhanced by adding the same amount of chitosan during the deposition. Figure 7.4 shows the scanning electron microscopy (SEM) images of the cotton fabric with ZnO and chitosan nanoparticles.

Seeded semi-continuous emulsion copolymerization was used for the synthesis of anti-microbial latexes with core shell structure [78]. The latex was based on anti-microbial macromonomer (GPHGH) and hydrophobic acrylate monomers. The polymerization was carried out in the presence of a cationic surfactant. In another study, covalent mechanism was used to synthesize anti-microbial L-cysteine (L-Cys)-functionalized cotton [79]. The covalently bound cotton threads showed superior antimicrobial capacities of 83% and 89% against S. aureus K. pneumoniae respectively.

Inorganic-organic hybrid polymers modified/filled with ZnO nanoparticles were attached to cotton/polyester (65/35%) and cotton (100%) fabrics by Farouk *et al.* [80]. The composite material showed superior ability to inhibit a wide range of bacterial infections. In another study, peppermint releasing alginate nanocapsules made by micro-emulsion method were applied to cotton fabrics using microwave curing method [81]. Various parameters were optimized to obtain a fabric with optimal anti-microbial properties. In optimum conditions, the fabric showed 100% reduction in the amount of *E. coli* and *S. aureus* bacteria. Even after 25 washing cycles, the composite were observed to retain 16% of the initial amount of peppermint oil. Composites containing chitosan (CH) and alginate (ALG) layers were prepared using layer by layer method and attached to cotton fabrics using electrostatic method [82]. The composites had bacteriostatic effects on bacteria, viz. *Klebsiella pneumonia* and Staphylococcus *aureus.* The composite with five multilayers (CH/ALG/CH/ALG/CH) was observed to be most effective for bacteriostatic inhibition.

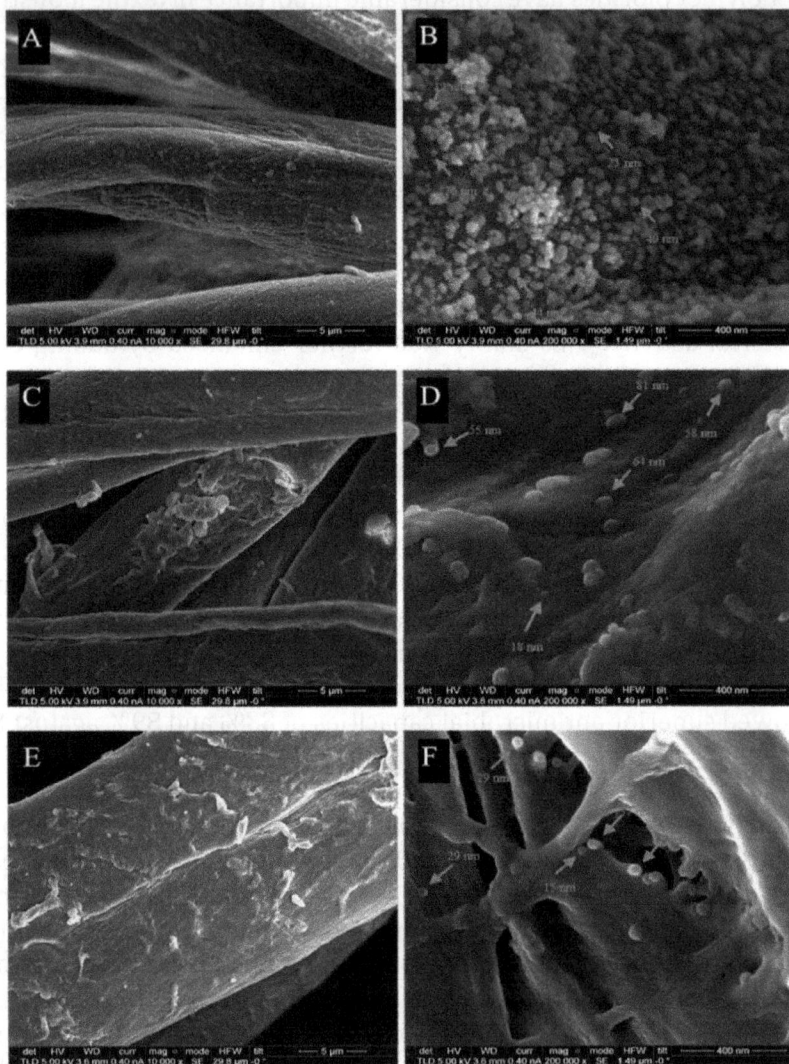

Figure 7.4 SEM images of cotton fabric coated with (A,B) ZnO, (C,D) ZnO/CS and (E,F) CS. Reproduced from Reference 77 with permission from American Chemical Society.

Contact lenses, hip implants, cardiac pacemakers, catheters, etc., have brought a tremendous change in the medical industry. The presence of biofilms in these implants creates issues due to various infections [83,84]. Microbial infection causes almost 20% mortalities worldwide, and 80% of these result from biofilm formation [85,86]. A range of organisms has been specifically implicated in device/bio-material-related infections [87-90]. It involves both fungal groups and bacteria [91,92]. New bifunctional amphiphilic random copolymers were developed for use in medical devices as coatings [93], by polymerization of an anti-microbial cationic monomer, antioxidant and anti-microbial hydrophobic monomer (containing hydroxytyrosol, HTy). The authors observed that the lowest minimal inhibitory concentration against Staphylococcus epidermidis was shown by the copolymer having the highest HTy molar content. In another study, PVC, PU and silicone were coated with anti-microbial nanoparticles [94]. Specifically, the *in-situ* coating of selenium (Se) nanoparticles rendered these polymers anti-bacterial abilities. The density of the Se atoms on the polymer surface had a direct influence on its anti-microbial activity. Polymer-peptide conjugates were also used for coating biomaterial implants and devices for infection resistance by Muszanska *et al.* (Figure 7.5) [95]. Anti-microbial peptides (AMP) were used to functionalize anti-adhesive polymer brushes made from block copolymer Pluronic F-127 (PF127). Adhesion and spreading of host tissue cells were promoted by arginine-glycine-aspartate (RGD) peptides. Without hampering tissue compatibility, the coatings composed of a suitable ratio of the functional constituents: PF127, PF127 modified with AMP and PF127 modified with RGD showed good anti-adhesive and bactericidal properties.

The insoluble disinfectants initiate the microbial inactivation mechanism by the interaction of the reactive species in the bulk phase. Polymeric disinfectants are an ideal choice for this application. [96-100]. In one such study, thiol functionalized PVDF membranes were developed with surface assembled silver nanoparticles [101]. The formation of the nanocluster assembly of silver nanoparticles was facilitated by the esterification reaction between alkaline treated PVDF membrane (TGA-PVDF) and thioglycolic acid (TGA). In another study, Ahmed *et al.* [102] reported the improved bactericidal behavior of the polyvinyl-*N*-carbazole (PVK) and of single-walled carbon nanotubes (SWNTs) (97:3 wt% ratio PVK:SWNTs) composite, as compared to 100% SWNTs coated membranes. The PVK-SWNTs membrane was completely non-toxic to fibroblast cells as opposed to pure SWNTs, which showed acute levels of toxicity to the exposed cell lines.

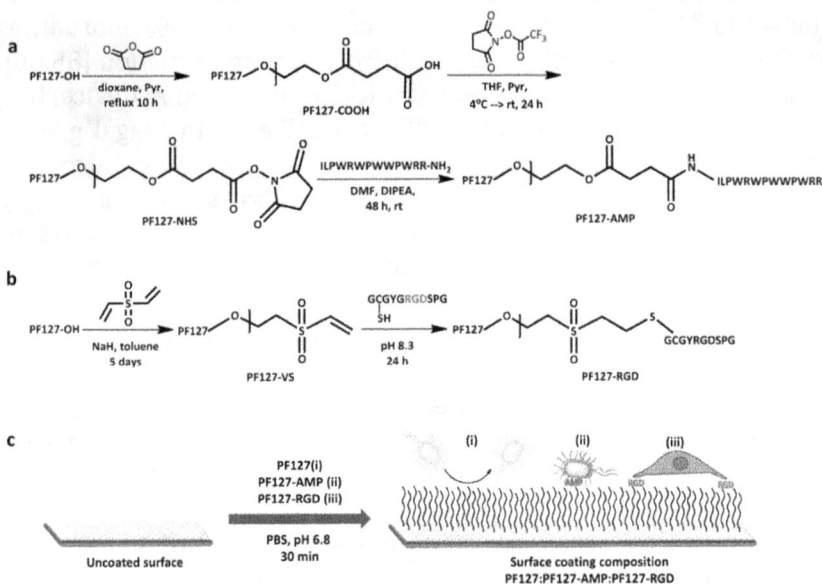

Figure 7.5 (a) PF127-AMP, (b) PF127-RGD and (c) the triple activity coating made by immobilizing the conjugates on a silicone rubber surface using dip-coating. Reproduced from Reference 95 with permission from American Chemical Society.

Bollmann *et al.* [103] studied polyacrylate-water partitioning of biocidal compounds. The polyacrylate-water partition constants were observed to be predominantly below octanol-water partition constants. Capsaicin-mimic materials were also reported by Zhang *et al.* [104] as anti-fouling membranes for water treatment. Polyaniline Th(IV) tungstomolybdophosphate (PANI/TWMP) nanocomposite ion exchanger was used as a biocidal membrane by Sharma *et al.* [105], which exhibited abilities to filter heavy metals from water.

Highly efficient phosphate scavenger to realize nutrient-starvation anti-bacteria was designed using La(OH)$_3$ nanorods immobilized in polyacrylonitrile (PAN) nanofibers (PLNFs). Electrospinning and a subsequent *in-situ* surfactant-free precipitation method were used to synthesize the composite. The La(OH)$_3$ nanorods with PAN protection showed 8 times more phosphate capture capacity than La(OH)$_3$ nanocrystals without PAN protection. Figure 7.6 illustrates the nutrient-starvation anti-bacteria by PLNFs.

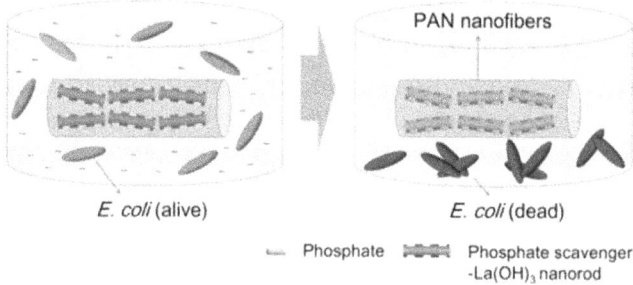

Figure 7.6 Schematic illustration of the nutrient-starvation anti-bacteria by PLNFs. Reproduced from Reference 106 with permission from American Chemical Society.

The idea of anti-microbial packaging has received significant attention in view of its capability to upgrade food safety [107]. Gemili *et al.* [108] incorporated cellulose acetate films with lysosome for obtaining anti-microbial packaging materials. By changing the composition of the initial casting solution, the structure of the films could be changed from highly asymmetric to densely porous, which, in turn, facilitated the controlled release of lysosome. Dutta *et al.* [109] also reviewed various preparative methods and anti-microbial activity of chitosan based films for food applications. In another study, lysosome was incorporated into various polymers, viz. cellulose triacetate (CTA) films, nylon 6,6 pellets, polyvinyl alcohol (PVOH) beads, etc. [110]. Against a suspension of dried *Micrococcus lysodeikticus* cells, polyvinyl alcohol and nylon 6,6 yielded low activity, while CTA yielded the highest activity for dried cells in 30 min. Jin and Zhang [111] reported nisin incorporated biodegradable polylactic acid (PLA) for use in anti-microbial food packaging. Liquid foods (liquid egg white and orange juice) and culture media were used for studying the anti-microbial activity of PLA/nisin films against Salmonella Enteritidis, Escherichia coli O157:H7 and Listeria monocytogenes. The nisin particles inhibited L. monocytogenes growth in the liquid egg white and culture media.

7.3 Conclusions

Modern society is faced with a number of infections, and their control is of high importance. Disinfection and anti-microbial surfaces are the main strategies to eradicate these infections. Development of resistant microbial strains and environmental pollution caused by the disinfectants make them less attractive for everyday use. On the other hand, anti-microbial coatings, which release biocides on coming in contact with microbes, present much effective alternatives. In this respect, the use of polymers and polymeric coatings with biocidal properties is very promising and is envisaged to evolve further in the coming years.

References

1. Chang, C. C., Lin, C. K., Chan, C. C., Hsu, C. S., and Chen, C. Y. (2006) Photocatalytic properties of nanocrystalline TiO_2 thin film with Ag additions. *Thin Solid Films*, **494**, 274-278.
2. Dastjerdi, R., Mojtahedi, M. R. M., and Shoshtari, A. M. (2008) Investigating the effect of various blend ratios of prepared masterbatch containing Ag/TiO2 nanocomposite on the properties of bioactive continuous filament yarns. *Fibers and Polymers*, **9**, 727-734.
3. Yuranova, T., Mosteo, R., Bandara, J., Laubb, D., and Kiwi, J. (2006) Self-cleaning cotton textiles surfaces modified by photoactive SiO_2/TiO_2 coating. *Journal of Molecular Catalysis A: Chemical*, **244**, 160-167.
4. Nersisyan, H. H., Lee, J. H., Son, H. T., Won, C. W., and Maeng, D. Y. (2003) A new and effective chemical reduction method for preparation of nanosized silver powder and colloid dispersion. *Materials Research Bulletin*, **38**, 949-956.
5. Yeo, S. Y., Lee, H. J., and Jeong, S. H. (2003) Preparation of nanocomposite fibers for permanent antibacterial effect. *Journal of Materials Science*, **38**, 2143-2147.
6. Xie, Y., Ye, R., and Liu, H. (2006) Synthesis of silver nanoparticles in reverse micelles stabilized by natural bio surfactant. *Colloids and Surfaces A: Physicochemical Engineering Aspects*, **279**, 175-178.
7. Shahverdi, A. R., Minaeian, S., Shahverdi, H. R., Jamalifar, H., and Nohi, A. A. (2007) Rapid synthesis of silver nanoparticles using culture supernatants of Enterobacteria: a novel biological approach. *Process Biochemistry*, **42**, 919-923.
8. Jeong, S. H., Hwang, Y. H., and Yi, S. C. (2005) Antibacterial properties of padded PP/PE nonwovens incorporating nano-sized silver colloids. *Journal of Materials Science*, **40**, 5413-5418.
9. Wright, T. (2002) Alphasan: A thermally stable silver-based inorganic antimicrobial technology. *Chemical Fiber International*, **52**, 125.
10. Kumar, V. S., Nagaraja, B. M., Shashikala, V., Padmasri, A. H., Madhavendra, S. S., and Raju, B. D. (2004) Highly efficient Ag/C catalyst prepared by electro-chemical deposition method in controlli-

ng microorganisms in water. *Journal of Molecular Catalysis A: Chemical*, **223**, 313-319.

11. Vigo, T. L. (1994) *Textile Processing and Properties*, 11th volume, Elsevier, USA.

12. Yamamoto, T., Uchida, S., Kurihara, Y., and Nakayama, I. (1994) Jpn patent 94-204681.

13. Kenawy, E. R., Abdel-Hay, F. I., El-Shanshoury, A. R., and El-Newehy, M. H. (1998) Biologically active polymers: synthesis and antimicrobial activity of modified glycidyl methacrylate polymers having a quaternary ammonium and phosphonium groups. *Journal of Controlled Release*, **50**, 145-152.

14. Eknoian, M. W., Worley, S. D., and Harris, J. M. (1998) New biocidal N-halamine-PEG polymers. *Journal of Bioactive and Compatible Polymers*, **13**, 136-145.

15. Kanazawa, A., Ikeda, T., and Endo, T. (1993) Polymeric phosphonium salts as a novel class of cationic biocides. III. Immobilization of phosphonium salts by surface photografting and antibacterial activity of the surface-treated polymer films. *Journal of Polymer Science, Part A: Polymer Chemistry*, **31**, 1467-1472.

16. Ilker, M. F., Nuesslein, K., Tew, G. N., and Coughlin, E. B. (2004) Tuning the hemolytic and antibacterial activities of amphiphilic polynorbornene derivatives. *Journal of the American Chemical Society*, **126**, 15870-15875.

17. (a) Gabriel, G. J., Madkour, A. E., Dabkowski, J. M., Nelson, C. F., Nusslein, K., Tew, G. N. (2008) Synthetic mimic of antimicrobial peptide with nonmembrane-disrupting antibacterial properties. *Bio Macromolecules*, **9**, 2980-2983; (b) Lienkamp, K., Madkour, A. E., Musante, A., Nelson, C. F., Nusslein, K., Tew, G. N. (2008) Antimicrobial polymers prepared by ROMP with unprecedented selectivity: a molecular construction kit approach. *Journal of American Chemical Society*, **130**, 9836-9843.

18. Zasloff, M. (2002) Antimicrobial peptides of multicellular organisms. *Nature*, **415**, 389-395.

19. Shai, Y. (1999) Mechanism of the binding, insertion and destabilization of phospholipid bilayer membranes by alpha-helical antimicrobial and cell non-selective membrane-lytic peptides. *Biochimica et Biophysica Acta (BBA) - Biomembranes*. **1462**, 55-70.

20. Mowery, B. P., Lee, S. E., Kissounko, D. A., Epand, R. F., Epand, R. M., Weisblum, B., Stahl, S. S., and Gellman, S. H. (2007) Mimicry of anti-microbial host-defense peptides by random copolymers. *Journal of American Chemical Society*, **129**, 15474-15476.

21. Gelman, M. A., Weisblum, B., Lynn, D. M., and Gellman, S. H. (2004) Biocidal activity of polystyrenes that are cationic by virtue of protonation. *Organic Letters*, **6**, 557-560.

22. Palermo, E. F., and Kuroda, K. (2009) Chemical structure of cationic

groups in amphiphilic polymethacrylates modulates the antimicrobial and hemolytic activities. *Biomacromolecules*, **10**, 1416-1428.

23. Palermo, E. F., Sovadinova, I., and Kuroda, K. (2009) Structural determinants of antimicrobial activity and biocompatibility in membrane-disrupting methacrylamide random copolymers. *Biomacromolecules*, **10**, 3098-3107.

24. Huttinger, K. J., Muller, H., and Bomar, M. T. (1982) Synthesis and effect of carrier-bound disinfectants. *Journal of Colloid Interface Science*, **88**, 274-285.

25. Ren, H., Du, Y., Su, Y., Guo, Y., Zhu, Z., and Dong, A. (2018) A review on recent achievements and current challenges in antibacterial electrospun N-halamines. *Colloid and Interface Science Communications*, **24**, 24-34.

26. Liang, J., Barnes, K., Akdag, A., Worley, S. D., Lee, J., Broughton, R. M., and Huang, T. S. (2007) Improved antimicrobial Siloxane. *Industrial & Engineering Chemistry Research*, **46**, 1861-1866.

27. Charville, G. W., Hetrick, E. M., Geer, C. B., and Schoenfisch, M. H. (2008) Reduced bacterial adhesion to fibrinogen-coated substrates via nitric oxide release. *Biomaterials*, **29**, 4039-4044.

28. Ferrazzano, G. F., Roberto, L., Amato, I., Cantile, T., Sangianantoni, G., and Ingenito, A. (2011) Antimicrobial properties of green tea extract against cariogenic microflora: an in vivo study. *Journal of Medicinal Food*, **14**, 907-911.

29. Timofeeva, L., and Kleshcheva, N. (2011) Antimicrobial polymers: mechanism of action, factors of activity, and applications. *Applied Microbiology and Biotechnology*, **89**, 475-492.

30. Gabriel, G. J., Maegerlein, J. A., Nelson, C. F., Dabkowski, J. M., Eren, T., Nüsslein, K., and Tew, G. N. (2009) Comparison of facially amphiphilic versus segregated monomers in the design of antibacterial copolymers. *Chemistry - A European Journal*, **15**, 433-439.

31. Qi, F., Qian, Y., Shao, N., Zhou, R., Zhang, S., Lu, Z., Zhou, M., Xie, J., Wei, T., Yu, Q., and Liu, R. (2019) Practical preparation of infection-resistant biomedical surfaces from antimicrobial β-peptide polymers. *ACS Applied Materials & Interfaces*, **11**(21), 18907-18913.

32. Gilbert, P., and McBain, A. J. (2003) An evaluation of the potential impact of the increased use of biocides within consumer products upon the prevalence of antibiotic resistance. *Clinical Microbial Reviews*, **16**, 189-208.

33. Gilbert, P., and Moore, L. E. (2005) Cationic antiseptics: diversity of action under a common epithet. *Journal of Applied Microbiology*, **99**, 703-715.

34. Grapski, J. A., and Cooper, S. L. (2001) Synthesis and characterization of non-leaching biocidal polyurethanes. *Biomaterials*, **22**, 2239-2246.

35. Haldar, J., An, D., Alvarez de Cienfuegos, L., Chen, J., and Klibanov, A.

M., (2006) Polymeric coatings that inactivate both influenza virus and pathogenic bacteria. *Proceedings of the National Academy of Sciences of USA*, **103**, 17667-17671.

36. Tew, G. N., Scott, R. W., Klein, M. L., and DeGrado, W. F. (2010) De novo design of antimicrobial foldamers and small molecules: from discovery to practical application. *Accounts of Chemical Research*, **43**, 30-39.

37. Sun, Y., and Sun, G. (2003), Novel refreshable N-halamine polymeric biocides: Grafting hydantoin-containing monomers onto high performance fibers by a continuous process. *Journal of Applied Polymer Science*, **88**, 1032-1039.

38. Nigmatullin, R., Gao, F., and Konovalova, V. J. (2008) Polymer-layered silicate nanocomposites in the design of antimicrobial materials. *Journal of Material Science*, **43**, 5728-5733.

39. Alamri, A., El-Newehy, M. H., and Al-Deyab, S. S. (2012) Biocidal polymers: synthesis and antimicrobial properties of benzaldehyde derivatives immobilized onto amine-terminated polyacrylonitrile. *Chemistry Central Journal*, **6**, 111.

40. Ikeda, T., Tazuke, S., and Watanabe, M. (1983) Interaction of biologically active molecules with phospholipid membranes, *Biochimica ET Biophysica Acta (BBA) - Biomembranes*, **735**, 380-386.

41. Gholami, H., Yeganeh, H., Gharibi, R., Jalilian, M., and Sorayya, M. (2015) Catalyst free-click polymerization: A versatile method for the preparation of soybean oil based poly1,2,3-triazoles as coatings with efficient biocidal activity and excellent cytocompatibility. *Polymer*, **62**, 94-108.

42. Zhao, Q., Li, J., Zhang, X., Li, Z., and Tang, Y. (2016) Cationic Oligo(thiophene ethynylene) with broad-spectrum and high antibacterial efficiency under white light and specific biocidal activity against S. aureus in Dark. *ACS Applied Material Interfaces*, **8**, 1019-1024.

43. Kawabata, N. (1992) Capture of microorganisms and viruses by pyridinium-type polymers and application to biotechnology and water-purification. *Progress in Polymer Science*, **17**, 1-34.

44. Dizman, B., Elasri, M. O., and Mathias, L. J. (2005) Synthesis, characterization, and antibacterial activities of novel methacrylate polymers containing norfloxacin. *Biomacromolecules*, **6**, 514-520.

45. Lawson, M. K. C., Shoemaker, R., Hoth, K. B., Bowman, C. N., and Anseth, K. S. (2009) Polymerizable vancomycin derivatives for bactericidal biomaterial surface modification: structure-function evaluation. *Biomacromolecules*, **10**, 2221-2234.

46. Waschinski, C. J., and Tiller, J. C., (2005) Poly(oxazoline)s with telechelic antimicrobial functions. *Biomacromolecules*, **6**, 235-243.

47. Turos, E., Shim, J. Y., Wang, Y., Greenhalgh, K., Reddy, G. S. K., Dickey, S., and Lim D. V. (2007) Antibiotic-conjugated polyacrylate nanoparticles: New opportunities for development of anti-MRSA agents.

Bioorganic & Medicinal Chemistry Letters, **17**, 53-56.

48. Tashiro, T. (2001) Antibacterial and bacterium adsorbing macro-molecules. *Macromolecular Materials and Engineering*, **286**, 63-87.

49. Nathan, A., Zalipsky, S., Ertel, S. I., Agathos, S. N., Yarmush, M. L., and Kohn, J. (1993) Copolymers of lysine and polyethylene glycol: a new family of functionalized drug carriers. *Bioconjugate Chemistry*, **4**, 54-62.

50. Kukharenko, O., Bardeau, J. F., Zaets, I., Ovcharenko, L., Tarasyuk, O., Porhyn, S., Mischenko, I., Vovk, A., Rogalsky, S., and Kozyrovska, N. (2014) Promising low cost antimicrobial composite material based on bacterial cellulose and poly hexamethylene guanidine hydro-chloride. *European Polymer Journal*, **60**, 247-254.

51. Timofeeva, L. M., Kleshcheva, N. A., and Shleeva, M. O. (2015) Non-quaternary poly (diallylammonium) polymers with different amine structure and their biocidal effect on Mycobacterium tuberculosis and Mycobacterium smegmatis. *Applied Microbiological Biotechnology*, **99**, 2557-2571.

52. El-Newehy, M. H., Kenawy, E. R., and Al-Deyab S.S., (2014) Biocidal polymers: preparation and antimicrobial assessment of immobilized onium salts onto modified chitosan. *International Journal of Polymeric Materials and Polymeric Biomaterials*, **63**, 758-766.

53. Zhang, J., Chen, Y. P., Miller, K. P., Ganewatta, M. S., Bam, M., Yan, Y., Nagarkatti, M., Decho, A. W., and Tang, C. (2014) Antimicrobial metallo polymers and their bioconjugates with conventional antibiotics against multidrug-resistant bacteria. *Journal of American Chemical Society*, **136**, 4873-4876.

54. Sellenet, P. H., Allison, B., Applegate, B. M., and Youngblood, J. P. (2007) Blood synergistic activity of hydrophilic modification in antibiotic polymers. *Biomacromolecules*, **8**(1), 19-23.

55. *World Health Statistics 2009*, World Health Organization (2009). Online: https://www.who.int/whosis/whostat/2009/en/ [accessed 19th June 2019].

56. Gabriel, G. J., Som, A., Madkour, A. E., Eren, T., Tew, G. N., (2007) Infectious disease: connecting innate immunity to biocidal polymers. *Materials Science and Engineering R: Reports*, **57**, 28-64.

57. Spielberg, B., Powers, J. H., Brass. E. P., Miller. L. G., and Edwards, Jr., J. E. (2004) Trends in antimicrobial drug development: Implications for the future. *Clinical Infectious Diseases*, **38**, 1279-1286.

58. Farzinfar, E., and Paydayesh, A. (2018) Investigation of polyvinyl alcohol nanocomposite hydrogels containing chitosan nanoparticles as wound dressing. *International Journal of Polymeric Materials and Polymeric Biomaterials*, **68**, 628-638.

59. Evans R. C., and Holmes, C. J. (1987) Effect of vancomycin hydro-chloride on Staphylococcus epidermidis biofilm associated with silicone elastomer. *Antimicrobial Agents and Chemotherapy*, **31**, 889-

894.

60. Abel, T., Cohen, J. L. I., Engel, R., Filshtinskaya, M., Melkonian, A., and Melkonian, K. (2002) Preparation and investigation of antibacterial carbohydrate-based surfaces. *Carbohydrate Research*, **337**, 2495-2499.

61. Ewald, A., Glückermann, S. K., Thull, R., and Gbureck, U. (2006) Antimicrobial titanium/silver PVD coatings on titanium. *Biomedical Engineering Online*, **5**, 22.

62. Barasch, A., Elad, S., Altman, A., Damato, K., and Epstein, J. (2006) Antimicrobials, mucosal coating agents, anesthetics, analgesics, and nutritional supplements for alimentary tract mucositis. *Support Care Cancer*, **14**, 528-532.

63. Gehring, J., Schleheck, D., Trepka, B., and Polarz, S. (2015) Mesoporous organosilica nanoparticles containing superacid and click functionalities leading to cooperativity in biocidal coatings. *ACS Applied Material Interfaces*, **7**, 1021-1029.

64. Iannazzo, D., Pistone, A., Visco, A., Galtieri, G., Giofrè, S. V., Romeo, R., Romeo, G., Cappello, S., Bonsignore, M., and Denaro, R. (2015) 1,2,3-Triazole/MWCNT conjugates as filler for gelcoat nanocomposites: new active antibiofouling coatings for marine application. *Materials Research Express*, **2**(11), 3.

65. Finlay, J. A., Schultz, M. P., Cone, G., Callow, M. E., and Callow, J. A. (2013) A novel biofilm channel for evaluating the adhesion of diatoms to non-biocidal coatings. *Biofouling*, **29**(4), 401-411.

66. Schultz, M. P., Walker, J. M., Steppe, C. N., and Flack, K. A. (2015) Impact of diatomaceous biofilms on the frictional drag of fouling-release coatings. *Biofouling*, **31**, 759-773.

67. Watson, M. G., Scardino, A. J., Zalizniak, L., and Shimeta, J. (2015) Colonization and succession of marine biofilm-dwelling ciliate assemblages on biocidal antifouling and fouling-release coatings in temperate Australia. *Biofouling*, **31**, 709-720.

68. Muthukrishnan, T., Abed, R. M., Dobretsov, S., Kidd, B., and Finnie, A. A. (2014) Long-term micro fouling on commercial biocidal fouling control coatings. *Biofouling*, **30**(10), 1155-1164.

69. Kenawy, E. R., Worley, S. D., and Broughton, R. (2007) The Chemistry and applications of antimicrobial polymers: A state-of-the-art review. *Biomacromolecules*, **8**(5), 1359-1384.

70. *Textile Finishing*, Heywood, D. (ed.), Society of Dyers and Colorists, UK (2003).

71. Chung, D. W. and Lim, J. C. (2009) Study on the effect of structure of polydimethylsiloxane grafted with polyethylene oxide on surface activities. *Colloids and Surfaces A: Physicochemical and Engineering Aspects*, **336**, 35-40.

72. Kim, D. W., Noh, S. T., Jo, B. W. (2006) Effect of salt and pH on surface active properties of comb rake-type polysiloxane surfactants. *Collo-*

ids and Surfaces A: Physicochemical and Engineering Aspects, **287**, 106-116.

73. Dastjerdi, R., Montazer, M., and Shahsavan, S. (2009) Incorporation of texturing and yarn surface modification using nano sized colloidal particles. *Chemical Biology - Fundamental Problems of Bionanotechnology*.

74. Hofmann, H., Fink-Petri A., and Salaklang J., (2009) Nanoparticles for Diagnostic and Therapeutic Applications: The Potential of Superparamagnetic Iron Oxide Nanoparticles (SPION). *Proceedings of the 6th International Conference on Biomedical Applications of Nanotechnology*, Germany.

75. Geppert M., Hohnholt M., Grunwald I., Baeumer M., Dringen R., (2009) Accumulation of iron oxide nanoparticles in cultured brain astrocytes. *Journal of Biomedical Nanotechnology*, **5**(3), 285-293.

76. De Santa Maria, L. C., Souza, J. D. C., Aguiar, M. R. M. P., Wang, S. H., Mazzei, J. L., Felzenszwalb, I., and Amico, S. C. (2008) Synthesis characterization, and bactericidal properties of composites based on crosslinked resins containing silver. *Journal of Applied Polymer Science*, **107**, 1879-1886.

77. Petkova, P., Francesko, A., Fernandes, M. M., Mendoza, E., Perelshtein, I., Gedanken, A., and Tzanov, T. (2014) Sonochemical coating of textiles with hybrid ZnO/chitosan antimicrobial nanoparticles. *ACS Applied Materials and Interfaces*, **6**(2), 1164-1172.

78. Pan, Y., Xiao, H., Cai, P., and Colpitts, M. (2016), Cellulose fibers modified with nano-sized antimicrobial polymer latex for pathogen deactivation. *Carbohydrate Polymers*, **135**, 94-100.

79. Nogueira F., Vaz, J., Mouro, C., Piskin, E., and Gouveia, I. (2014) Covalent modification of cellulosic-based textiles: A new strategy to obtain antimicrobial properties. *Biotechnology and Bioprocess Engineering*, **19**(3), 526-533.

80. Farouk, A., Moussa, S., Ulbricht, M., Schollmeyer, E., and Textor, T. (2013) ZnO-modified hybrid polymers as an antibacterial finish for textiles. *Textile Research Journal*, **84**(1), 40-51.

81. Soraya, G., and Mortazavi, S. M. (2015) Microwave curing for applying polymeric nanocapsules containing essential oils on cotton fabric to produce antimicrobial and fragrant textiles, *Cellulose*, **22**(6), 4065-4075.

82. Gomes, A. P., Mano, J. F., Queiroz, J. A., and Gouveia, I. C. (2013) Layer-by-layer deposition of antimicrobial polymers on cellulosic fibers: a new strategy to develop bioactive textiles, *Polymers for Advanced Technologies*, **24**, 1005-1010.

83. Lynch, A. S., and Robertson, G. T. (2008), Bacterial and fungal biofilm infections. *Annual Review of Medicine*, **59**, 415-428.

84. Busscher, H. J., van der Mei, H. C., Subbiahdoss, G., Jutte, P. C., van den Dungen, J. A. M., Zaat, S. A. J., Schultz, M. J., and Grainger, D. W.

(2012) Biomaterial-associated infection: locating the finish line in the race for the surface. *Science Translation Medicine*, **4**(153), 110.

85. *The World Health Report 2004: Changing History*, World Health Organization (2004). Online: https://www.who.int/whr/2004/en/ [accessed 21st June 2019].

86. Boucher, H. W., Talbot, G. H., Bradley, J. S., Edwards, J. E., Gilbert, D., Rice, L. B., Scheld, M., Spellberg, B., and J., Bartlett (2009) Bad bugs, no drugs: No ESKAPE! An update from the Infectious Diseases Society of America. *Clinical Infectious Diseases*, **48**, 1-12.

87. Martinez, L. R., and Casadevall, A., (2007), Cryptococcus neoformans biofilm formation depends on surface support and carbon source and reduces fungal cell susceptibility to heat, cold, and UV light. *Applied Environmental Microbiology*, **73**, 4592-4601.

88. Shingu-Vazquez, M., and Traven, A. (2011), Mitochondria and fungal pathogenesis: drug tolerance, virulence, and potential for antifungal therapy. *Eukaryotic Cell, American Society of Microbiology*, **10**, 1376-1383.

89. Finkel, J. S., and Mitchell, A. P. (2011) Mitchell Genetic control of Candida albicans biofilm development. *Nature Reviews Microbiology*, **9**, 109-118.

90. Rodriguez-Emmenegger, C., Brynda, E., Riedel, T., Houska, M., Šubr, V., Alles, A. B., Hasan, E., Gautrot, J. E., and Huck, W. T. S. (2011) Polymer brushes showing non-fouling in blood plasma challenge the currently accepted design of protein resistant surfaces. *Macromolecular Rapid Communication*, **32**, 952-957.

91. *The Direct Medical Costs of Healthcare-Associated Infections in U.S. Hospitals and the Benefits of Prevention* (2009). Online: https://stacks.cdc.gov/view/cdc/11550 [accessed 21st June 2019].

92. Singh, B. (2012) Human pathogens utilize host extracellular matrix proteins laminin and collagen for adhesion and invasion of the host. *FEMS Microbiology Reviews*, **36**, 1122-1180.

93. Taresco, V., Crisante, F., Francolini, I., Martinelli, A., D'Ilario, L., Ricci-Vitiani, L., Buccarelli, M., Pietrelli, L., and Piozzi, A. (2015) Antimicrobial and antioxidant amphiphilic random copolymers to address medical device-centered infections. *Acta Biomaterialia*, **22**, 131-140.

94. Tran, P. A., and Webster, T. J. (2013) Antimicrobial selenium nanoparticle coatings on polymeric medical devices. *Nanotechnology*, **24**, 155101.

95. Muszanska, A. K., Rochford, E. T. J., Gruszka, A., Gruszka, A., Andreas, A. B., Henk, J. B., Willem, N., Henny, C. V., and Andreas, H. (2014) Antiadhesive polymer brush coating functionalized with antimicrobial and RGD peptides to reduce biofilm formation and enhance tissue integration. *Biomacromolecules*, **15**, 2019-2026.

96. Tyagi, M., and Singh, H. J. (2000) Iodinated P (MMA-NVP): An effi-

cient matrix for disinfection of water. *Journal of Applied Polymer Science*, **76**, 1109-1116.

97. Tan, S., Li, G., Shen, J., Liu, Y., and Zong, M. (2000) Study of modified polypropylene nonwoven cloth. II. Antibacterial activity of modified polypropylene nonwoven cloths. *Journal of Applied Polymer Science*, **77**(9), 1869-1876.

98. Sun, Y., and Sun, G. (2001), Durable and refreshable polymeric N-halamine biocides containing 3-(4'-vinylbenzyl)-5, 5-dimethylhydantoin. *Journal of Polymer Science, Part A: Polymer Chemistry*, **39**, 3348-3355.

99. Chen, Y., Worley, S. D., Kim, J., Wei, C.-I., Chen, T.-Y., Santiago, J. I., Williams, J. F., and Sun, G. (2003) Biocidal Poly(styrenehydantoin) beads for disinfection of water. *Industrial & Engineering Chemistry Research*, **42**, 280-284.

100. Duncan, R., and Kopeček, J. (1984) Soluble synthetic polymers as potential drug carriers. *Advances in Polymer Science*, **57**, 51-101.

101. Sharma, M., Padmavathy, N., Remanan, S., Madras, G., and Bose, S. (2016) Facile one-pot scalable strategy to engineer biocidal silver nanocluster assembly on thiolated PVDF membranes for water purification. *RSC Advances*, **6**, 38972-38983.

102. Ahmed, F., Santos, C. M., Mangadlao, J., Advincula, R., and Rodrigues, D. F. (2013) Antimicrobial PVK: SWNT nanocomposite coated membrane for water purification: Performance and toxicity testing. *Water Research*, **47**(12), 3966-3975.

103. Bollmann, U. E., Ou, Y., Mayer, P., Trapp, S., and Bester, K. (2015) Polyacrylate–water partitioning of biocidal compounds: Enhancing the understanding of biocide partitioning between render and water. *Chemosphere*, **119**, 1021-1026.

104. Zhang, L., Xu, J., Tang, Y., Hou, J., Yu, L., and Gao, C. (2016) A novel long-lasting antifouling membrane modified with bifunctional capsaicin-mimic moieties via in situ polymerization for efficient water purification. *Journal of Materials Chemistry A*, **4**, 10352-10362.

105. Sharma, G., Pathania, D., Naushad, M., and Kothiyal, N.C. (2014) Fabrication, characterization and antimicrobial activity of polyaniline Th(IV) tungstomolybdophosphate nanocomposite material: Efficient removal of toxic metal ions from water. *Chemical Engineering Journal*, **251**, 413-421.

106. He, J., Wang, W., Sun, F., Shi, W., Qi, D., Wang, K., Shi,   R., Cui, F., Wang, C., and Chen, X. (2015) Highly efficient phosphate scavenger based on well-dispersed la(oh)3 nanorods in polyacrylonitrile nanofibers for nutrient-starvation antibacteria. *ACS Nano*, **9**(9), 9292-9302.

107. Bastarrachea, L., Dhawan, S., and Sablani, S. S. (2011) Engineering properties of polymeric-based antimicrobial films for food packaging: A review. *Food Engineering Reviews*, **3**, 79.

108. Gemili, S., Yemenicioglu, A., and Altinkaya, S. A. (2009) Development of cellulose acetate based antimicrobial food packaging materials for controlled release of lysozyme. *Journal of Food Engineering*, **90**(4), 453-462.
109. Dutta, P. K., Tripathi, S., Mehrotra, G. K., and Dutta, J. (2009) Perspectives for chitosan based antimicrobial films in food applications. *Food Chemistry*, **114**(4), 1173-1182.
110. Appendini, P., and Hotchkiss, J. H. (1997) Immobilization of lysozyme on food contact polymers as potential antimicrobial films. *Packaging Technology and Science*, **10**, 271-279.
111. Jin, T., and Zhang, H. (2008) Biodegradable polylactic acid polymer with nisin for use in antimicrobial food packaging. *Journal of Food Science*, **73**, M127-M134.

8

Living Polymer Architectures

8.1 Introduction

Living polymerization, also known as controlled polymerization, has become an important approach for the preparation of synthetic polymers having complex structures and controlled molecular weight distribution. Out of several living polymerization techniques, stable free radical polymerization, nitroxide mediated polymerization (NMP), reversible addition fragmentation chain transfer (RAFT), atom transfer radical polymerization (ATRP), etc., have proven to be the most noticeable methods [1,2].

8.2 Living/Controlled Polymerization Techniques

In controlled polymerization, there is a possibility to reach a point of control on the resulting polymer, thus, achieving a required molecular weight with narrow distribution, if we are able to control the transfer of chain and chain termination during polymerization [3-5]. In such a scenario, the number of initiator groups equate to the number of polymer chains, thus, the process continues as long as the monomer is entirely consumed. The ends of the polymer chains stay active, and, hence, the polymerization can be continued further by adding more monomers [6-9].

8.2.1 Nitroxide Mediated Polymerization

This polymerization technique needs the addition of appropriate alkoxyamine to the polymerization process and has drawn a significant research interest owing to its simplicity. This method is based on the dissociation of dormant species and initiating alkoxyamine [10,11] reversibly, which avoids irreversible termination. Therefore, most of the dormant living chains may grow as long as the monomer

Swati Singh and Vikas Mittal*, The Petroleum Institute (part of Khalifa University of Science and Technology), Abu Dhabi, UAE
*Current address: Bletchington, Wellington County, Australia

molecules are available, thus, resulting in narrow molar mass distribution. It was first reported by Rizzardo [12] and Georges [13] that 2,2,6,6-tetramethyl-1-piperidynyloxy nitroxide (TEMPO) can be used to control the polymerization of styrene. A large variety of nitroxides has been developed in order to expand the use for various monomers [14-22]. Boonpangrak *et al.* [23] also reported the synthesis of molecularly imprinted polymers (MIPs) via NMP. In the study, cholesterol-imprinted divinylbenzene (DVB)-based bulk MIPs were synthesized at 125 °C using cholesteryl(4-vinyl)phenyl carbonate as a template/functional monomer complex. Upon hydrolysis and removal of the template, the radio ligand binding experiments showed that the imprinted matrix obtained by NMP had an improved imprinting factor and binding affinity with respect to equivalent MIPs synthesized by free radical polymerization using benzoyl peroxide. Tomoeda *et al.* [24] polymerized n-butyl acrylate (BA) using emulsion nitroxide mediated polymerization. In this process at 100 °C, the authors used tetradecyltrimethylammonium bromide based cationic emulsifier and N-tert-butyl-N-(1-diethyl phosphono-2,2-dimethylpropyl) nitroxide. Molecular weight distribution of the polymers has been depicted in Figure 8.1.

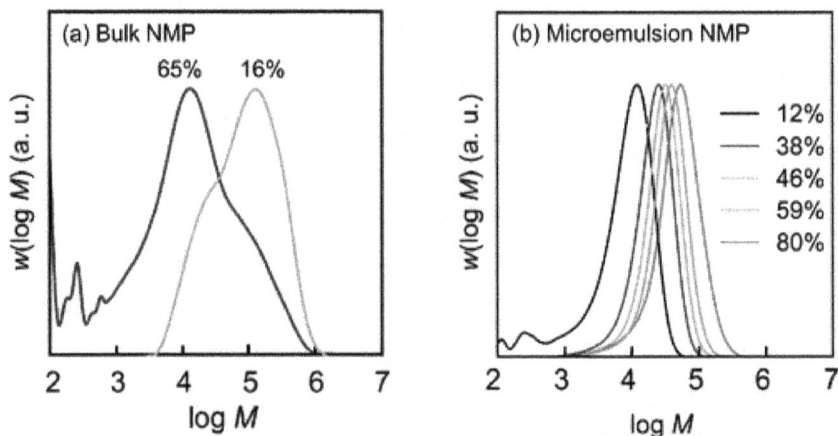

Figure 8.1 Molecular weight distributions (MWDs) at various conversions for (a) bulk (conversions: 16, 65%) and (b) microemulsion (conversions: 12, 38, 46, 59, 80%) NMPs of BA using SG1 at 100 °C: [AIBN]$_0$= 18.6mmol/L-monomer; [SG1]$_0$/[AIBN]$_0$ = 1.68, molar ratio. Reproduced from Reference 24 with permission from American Chemical Society.

8.2.2 Living Anionic and Telluride Mediated Polymerizations

Branched and linear polydiene structures can be synthesized using living anionic polymerization [25-30]. On the other hand, the living radical polymerization using organotellurium (TERP) can be used to synthesize polyacrylates [30-36]. Owing to the sensitivity of organomonotellurium compounds towards oxygen, it is important to provide inert atmosphere during synthesis. Azoinitiator and organoditelluride (DT) have been used by Yamago *et al.* [38], and DT was reported to be more stable than organomonotellurium compounds in ambient conditions [37,38]. Benedikt *et al.* [39] demonstrated telluride mediated polymerization method using benzoyl phenyltelluride as the reagent (Figure 8.2). It was observed that the material provided a low polydispersity in the range of 1.2-1.3. Mishima and Yamago [40] prepared copolymers comprising of methacrylates and α-alkenes using TERP technique. It was observed that the copolymers had a low polydispersity of less than 1.4. Other studies have reported similar use of TERP for generating living architectures [41-44].

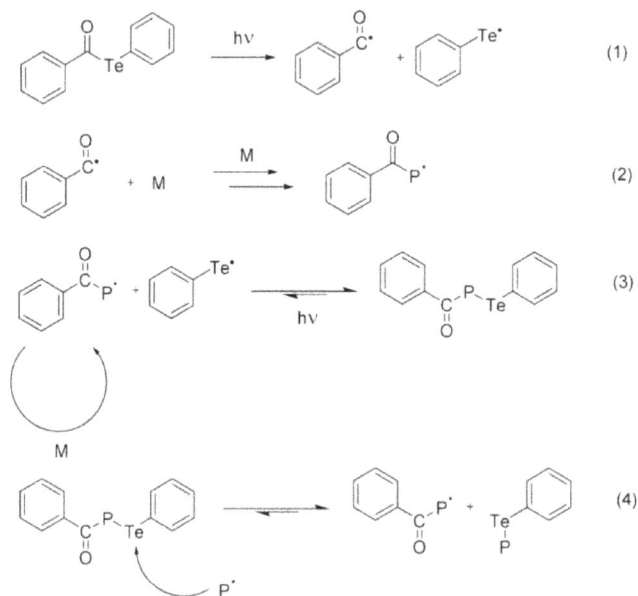

Figure 8.2 Schematic of living radical polymerization with benzoyl phenyltelluride as TERP-reagent. Reproduced from Reference 39 with permission from American Chemical Society.

8.2.3 Reversible Addition Fragmentation Chain Transfer

RAFT process combines the use of a chain transfer agent for achieving living structures [45-47]. Dithioesters [45] dithiocarbamates [48,49], trithiocarbonates [50] and xanthates [48,51] are the most widely used RAFT agents. Polymers with enhanced functionality and lower polydispersity index can be synthesized by choosing the suitable RAFT agent. For biomedical applications, RAFT polymers synthesized using biologically friendly solvents have also been explored. For instance, Ladavière *et al.* [52] performed the controlled polymerization of acrylic acid (AA) in alcohol and water, especially with phenoxyxanthates or with trithiocarbonate. Another study focused on the RAFT synthesis of poly(acrylic acid) in water with 4,4'-azobis(4-cyanovaleric acid) at 65 °C in the presence of water-soluble trithiocarbonate RAFT agent [53]. The resulting polymers were observed to have low dispersity (≤ 1.2).

Ionizing radiation can be a method of choice for additive-free (i.e. initiator, accelerator, catalyst, etc.) synthesis of polymers at room temperature [54]. Hong *et al.* [55] used 60Co γ-rays for the initiation of the RAFT polymerization of AA and synthesized PAA with controlled molecular weight (M_n=4100-8800 g/mol) and dispersities (1.07-1.22). Millard *et al.* [56] obtained PAA with excellent control and without irreversible termination even at high conversion and high molecular weights (up to 143,000 g/mol) in the presence of S,S-bis(α,α'-dimethyl-α''-acetic acid)trithiocarbonate (TRITT) and 3-benzylsulfanylthiocarbonylsulfanyl propionic acid (BPATT) RAFT agents. Subsequently, the living character of PAA was confirmed by using it as a macrotransfer agent to synthesize block copolymers with acrylamide (AA_m), N-isopropylacrylamide (NIPAAm) and N,N-dimethylacrylamide (DMAAm) under γ-irradiation in aqueous media [57]. Klimkevicius and Makuska [58] reported successive RAFT polymerization of poly(ethylene oxide) methyl ether methacrylates with different length of PEO chains. Kulai *et al.* [59] described a novel array RAFT agents based on tin and evaluated the polymerization of styrene, methyl acrylate and acrylamides. ^{119}Sn NMR was reported to be a helpful tool to control the RAFT polymerization mediated by Sn (Figure 8.3). Moreover, it was observed that Sn-RAFT agents were highly reactive as anticipated from the electron donating triphenylstannyl analogs. This drove the RAFT polymerization of more activated monomers, but completely prevented the polymerization of less activated monomers.

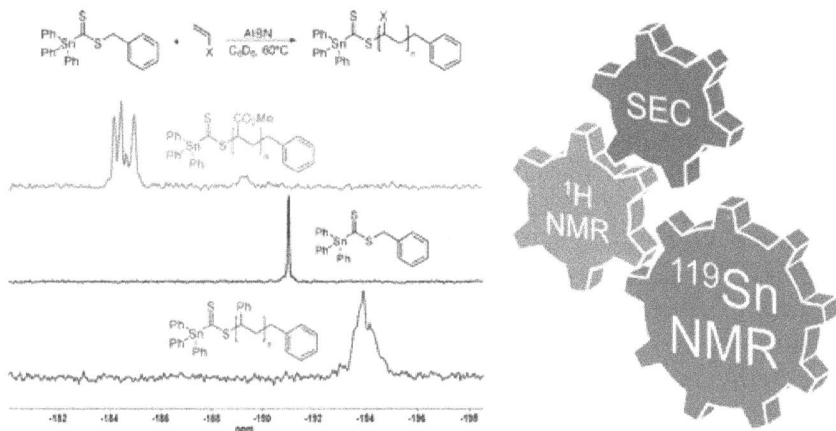

Figure 8.3 Schematic showing the use of ^{119}Sn NMR in RAFT polymerization. Reproduced from Reference 59 with permission from American Chemical Society.

Chaduc *et al.* [60] also studied the RAFT polymerization of methacyrlic acid in water. The polymer chains displayed a low dispersity (<1.19) in various conditions. Zhou *et al.* [61] studied seeded dispersion RAFT polymerization, poly(ethylene glycol) macro-RAFT agent mediated dispersion RAFT polymerization and seeded emulsion RAFT polymerization. It was observed that the dispersion RAFT polymerization proceeded slowly as compared to the other two. Also, the dispersion RAFT polymerization afforded vesicles, seeded dispersion RAFT polymerization led to the mixture of vesicles and porous nanospheres, whereas the seeded emulsion RAFT polymerization afforded porous nanospheres. Nicolay [62] also prepared copolymers of polythiol using RAFT polymerization method. In another study, Grover *et al.* [63] reported the preparation of polymers based on poly(PEGA) with pyridyl disulfide groups or aminooxy groups at the end of the chains by RAFT polymerization.

8.2.4 Atom Transfer Radical Polymerization

This polymerization technique is based on the transition metal facilitated atom transfer radical addition reactions [64-66]. The formation of propagating radicals is realized by the initiator, and a lower oxidation state transition metallic complex with halide coordinate [67] can terminate the process. Persistent radical effect can result in the reduction of termination as the reaction is progressing [68,69]. This results in increased chain length, viscosity and conversion [70]. As a consequence, there is a shift in the equilibrium towards the dormant species [69]. This process can be considered to be complex as it involves more than two oxidation states and a transition metal complex [71]. Equilibrium can be formed because of the different interactions between the species in the reaction medium [72,73]. The initiators usually used in the process are alkyl (pseudo)halides, which can be high or low molar mass compounds.

Shipp *et al.* [74] reported the preparation of copolymers containing methyl acrylate, methyl methacrylate (MMA) and butyl acrylate through atom transfer radical polymerization based on copper. A difunctional center block was developed following the addition of MMA to the two ends during the formation of PMMA-b-PBA-b-PMMA triblock copolymer. Figure 8.4 shows the molecular distribution of Br-PMA-Br macroinitiator and PMMA-b-PMA-b-PMMA

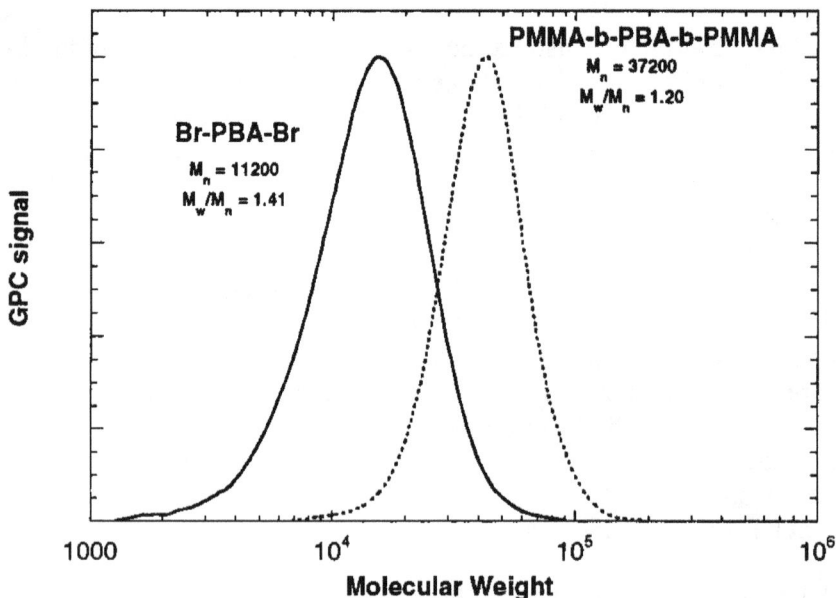

Figure 8.4 Molecular weight distributions of Br-PMA-Br macroinitiator and PMMA-b-PMA-b-PMMA triblock copolymer. Reproduced from Reference 74 with permission from American Chemical Society.

triblock copolymer. In another study, Robinson *et al.* [75] synthesized 2-hydroxyethyl methacrylate using ATRP at ambient temperature. ATRP and RAFT were combined for the synthesis of poly(vinyl acetate)-b-polystyrene (PVAc-b-PSt) with a well-defined structure by Tong *et al.* [76]. Successful implementation of activators regenerated by electron transfer atom transfer radical polymerization (ARGET ATRP) for aqueous media was achieved for the first time by Simakova *et al.* [77].

8.3 Advantages of Living Polymerization Techniques

Out of the various living polymerization techniques, RAFT offers some unique benefits, which include the capability to control the molecular weight and molecular weight distribution as well as the ability to resume the polymerization using dormant materials [78,79]. The various chain transfer agents used in RAFT polymerization provide increased acceptance towards different solvents and monomers [79,80-86]. Atom-transfer radical-polymerization is also known as a flexible living polymerization method, and it is possible to use this method for a wide range of monomers under suitable conditions [86,87]. However, this method is less appropriate for the monomers having acrylic groups and low reactivity [87,88]. The formation of graft and block copolymers is also effectively achieved in this method by forming halogenated polymer chains towards the end of the ATRP polymerization process [89,90]. The removal of metal catalysts is, however, tough after the ATRP process [91]. NMP has proved to be advantageous in many aspects even though the method is not suitable for situations requiring high temperature or long reaction time [92]. It is also not ideal for vinyl acetate based monomers [93]. The most significant benefit of NMP is the capability to generate highly pure polymers [93]. Also, NMP has been proven to be highly useful in the synthesis of polymers using 1,3-dienes [93]. NMP is also particularly valuable for the preparation of styrene [94,95].

8.4 Applications of Living Polymer Architectures

Living radical polymerization has driven a significant research interest in recent years owing to the potential to achieve "smart" multifunctional drug delivery methods. The function of proteins can be modified using beneficial gene delivery by replacing the genes which are not functional. An array of therapeutics based on nucleic acids have been observed, which can lead to possible treatments for infectious, hereditary and acquired syndromes. Plasmids are present in the candidate drugs which encourage small interfering RNAs and gene expression, which subsequently calms the target genes. Introduction of living free radical polymerization has brought an improvement in the control, quality and reproducibility of such materials. It is, thus, possible to achieve materials with required molecular weights and structures, along with narrow dispersion. Consequently, the impact of structure and molecular weight of the polymers on cytotoxicity has been analyzed, which is helpful to develop the scheme of next-generation vectors [96].

A control over density, chemical configuration, architecture and size make the polymer brushes an interesting topic of research. A layer of polymer attached to one side of the substrate surface and the other end hanging into solvent make the polymer brushes. The polymer chains stretch away from the surface because of both osmotic and steric repulsion between the chain segments in an appropriate solvent. Generally utilized artificial approach for developing polymer brushes is surface-initiated polymerization. Surface initiated ATRP offers the synthesis of precise polymer brushes owing to the benefits including simple preparation of initiator, good regulation of the chain growth, polymerization in aqueous solution and "living" end for copolymer embedding [97,98]. Other polymerization approaches explored to produce polymer brushes include RAFT [99], ring-opening metathesis polymerization [100] and NMP [101]. Also, the very precise control of the composition and construction due to living polymerization allows the improvement of various advanced materials with specific properties suitable for targeted applications, such as separations (chromatography and membrane), corrosion resistance, drug delivery, cell growth control, etc. Polymer brushes are also of particular interest in medical applications such as biomaterial implants and medical devices as well as diagnostics, packaging and blood contact materials [102,103]. Functional polymer brushes, especially the patterned and gradient architectures, are also employed extensively for biological applications, such as biosensing and biomolecule immobilization [104,105]. Polymer brush-assisted electroless deposition of metals (e.g., Cu, Ni, Ag or Au) on various substrates combined with lithography or printing techniques allows one to prepare inorganic/organic hybrid materials with potential applications in flexible electronics [106-108]. Applications in solar cells, memory storage and sensors have also been indicated [97].

8.5 Conclusion

In this chapter, different living polymerization techniques have been briefly reviewed. The multitude of controlled/living polymerization methods has helped to develop polymers with tuned structural characteristics. Owing to this, the developed polymers exhibit "living" nature and are useful for a wide variety of innovative applications.

References

1. Hasirci, V., Huri, P. Y, Tanir, T. E., Eke, G., Hasirci, N. (2017) Polymer fundamentals: polymer synthesis. *Comprehensive Biomaterials*, **1**, 349-371.
2. Piotti, M. E. (2001) Living cationic polymerization. In: *Encyclopedia of Materials: Science and Technology*, Elsevier, USA.
3. Calhoun, A., and Peacock, A. (2006) Polymer *Chemistry: Properties and Applications*, Hanser, Germany.
4. *High Modulus Polymers*, Zachariades A. E., and Porter R. S. (eds.), Marcel Dekker Inc., USA (1988).
5. Odian, G. (1991) *Principles of Polymerization*, 3rd edition, Wiley, USA.
6. Szwarc, M. (1998) Living polymers. Their discovery, characterization, and properties. *Journal of Polymer Science, Part A: Polymer Chemistry*, **36**, 9-15.
7. Szwarc, M. (1956) 'Living' polymers. *Nature*, **178**, 1168-1169.
8. Xu, J., Jung, K., Atme, A., Shanmugam, S., and Boyer, C. (2014) A Robust and versatile photo induced living polymerization of conjugated and unconjugated monomers and its oxygen tolerance. *Journal of American Chemical Society*, **136**, 5508-5519.
9. Flory, P. (1953) *Principles of Polymer Chemistry*, Cornell University Press, USA.
10. He, J., Knoll, K., and McKee, G. (2004) Synthesis and properties of block copolymers of isoprene and 1,3-cyclohexadiene. *Macromolecules*, **37**, 4399-4405.
11. Goto, A., and Fukuda, T. (2004) Kinetics of living radical polymerization. *Progress in Polymer Science*, **29**, 329-385.
12. Solomon, D. H., Rizzardo, E., and Cacioli, P. (1986) Polymerization Process and Polymers Produced Thereby, patent US4581429.
13. Georges, M. K., Veregin, R. P. N., Kazmaier, P. M., and Hamer, G. K. (1993) Narrow molecular weight resins by a free-radical polymerization process. *Macromolecules*, **26**, 2987-2988.
14. Hong, K., and Mays, J. W. (2001) 1,3-Cyclohexadiene polymers. 1.

Anionic polymerization. *Macromolecules*, **34**, 782-786.

15. Grimaldi, S., Finet, J. P., Zeghdaoui, A., Tordo, P., Benoit, D., Gnanou, Y., Fontanille, M., Nicol, P., Pierson, J. F. (1997) *Polymer Preprints*, **38**(1), 651-652.

16. Benoit, D., Chaplinski, V., Braslau, R., Hawker, C. J. (1999) Development of a universal alkoxyamine for "living" free radical polymerizations. *Journal of American Chemical Society*, **121**, 3904-3920.

17. Benoit, D., Grimaldi, S., Robin, S., Finet, J. P., Tordo, P., and Gnanou, Y. (2000) Kinetics and mechanism of controlled free-radical polymerization of styrene and n-butyl acrylate in the presence of an acyclic β-phosphonylated nitroxide. *Journal of American Chemical Society*, **122**, 5929-5939.

18. Grimaldi, S., Lemoigne, F., Finet, J. P., Tordo, P., Nicol, P., and Plechot, M. (1996) Polymerization in the Presence of a β-Substituted Nitroxide Radical, patent WO1996024620A1.

19. Matyjaszewski, K. (2003) Controlled/living radical polymerization: State of the art in 2002. *ACS Symposium Series*, **854**, 2-9.

20. Sciannamea, V., Jerome, R., and Detrembleur, C. (2008) In-situ nitroxide-mediated radical polymerization (NMP) processes: their understanding and optimization. *Chemical Reviews*, **108**, 1104-1126.

21. Ide, N., and Fukuda, T. (1997) Nitroxide-controlled free-radical copolymerization of vinyl and divinyl monomers. Evaluation of pendant-vinyl reactivity. *Macromolecules*, **30**, 4268-4271.

22. Ide, N., and Fukuda, T. (1999) Nitroxide-controlled free-radical copolymerization of vinyl and divinyl monomers. 2. Gelation. *Macromolecules*, **32**, 95-99.

23. Boonpangrak, S., Whitcombe, M. J., Prachayasittikul, V., Mosbach, K., and Ye, L. (2006) Preparation of molecularly imprinted polymers using nitroxide-mediated living radical polymerization. *Biosensors and Bioelectronics*, **22**, 349-354.

24. Tomoeda, S., Kitayama, Y., Wakamatsu, J., Minami, H., Zetterlund, P. B., and Okubo M. (2011) Nitroxide-mediated radical polymerization in microemulsion (microemulsion NMP) of n-Butyl acrylate. *Macromolecules*, **44**, 5599-5604.

25. Shingo, K., Cheng, L., Thomas, R. H., and Marc, A. H. (2009) Controlled polymerization of a cyclic diene prepared from the ring-closing metathesis of a naturally occurring monoterpene. *Journal of American Chemical Society*, **131**, 7960-7961.

26. Hirao, A., Goseki, R., and Ishizone, T. (2014) Advances in living anionic polymerization: from functional monomers, polymerization systems, to macromolecular architectures. *Macromolecules*, **47**, 1883-1905.

27. Suzuki, T., Tsuji, Y., Takegami, Y., and Harwood, H. J. (1979) Micro-

structure of Poly(2-phenylbutadiene) prepared by anionic initiators. *Macromolecules*, **12**, 234-239.

28. (a) Inoue, H. I., Helbig, M., Vogl, O. (1977) Preparation and characterization of head-to-head polymers. 5. head-to-head polystyrene. *Macromolecules*, **10**, 1331-1339; (b) Akira, H., Yashunori, S., and Katsuhiko, T. (1998) Anionic living polymerization of 2,3-Diphenyl-1,3-butadiene. *Macromolecules*, **31**, 9141-9145; (c) Zhang, Y., Li, J., Li, X. H., He, J. P. (2014) Regio-specific polyacetylenes synthesized from anionic polymerizations of template monomers. *Macromolecules*, **47**, 6260-6269.

29. (a) Yuki, K., Keita, K., Sotaro, I., and Takashi, I. (2013) Living anionic polymerization of benzofulvene: highly reactive fixed transoid 1,3-Diene. *ACS Macro Letters*, **2**, 164-167; (b) Natori, I. (1997) Synthesis of polymers with an alicyclic structure in the main chain. living anionic polymerization of 1,3-cyclohexadiene with the n-butyllithium/n,n,n',n'-tetramethyl- ethylenediamine system. *Macromolecules*, **30**, 3696-3697.

30. Liu, K., He, Q., Ren, L., Gong, L. J., Hu, J. L., Ou, E. C., and Xu, W. J. (2016) Synthesis and characterization of the well-defined poly pentadiene via living anionic polymerization of (E)-1,3-pentadiene. *Polymer*, **89**, 28-40.

31. Yamago, S. (2006) Development of organotellurium-mediated and organostibine-mediated living radical polymerization reactions. *Journal of Polymer Science, Part A, Polymer Chemistry*, **44**, 1-12.

32. Yamago, S. (2009) Precision polymer synthesis by degenerative transfer controlled/living radical polymerization using organo tellurium, organostibine, and organobismuthine chain-transfer agents. *Chemical Reviews*, **109**, 5051-5068.

33. *Polymer Science: A Comprehensive Reference*, Matyjaszewski K., and Moller, M. (eds.), Elsevier, The Netherlands (2012).

34. Nakamura, Y., and Yamago, S. (2013) Organotellurium-mediated living radical polymerization under photoirradiation by a low-intensity light-emitting diode. *Polymer*, **54**, 981-994.

35. Zetterlund, P. B., Kagawa, Y., Okubo, M. (2008) Controlled/living radical polymerization in dispersed systems. *Chemical Reviews*, **108**, 3747-3794.

36. Huster, D., Yao, X., and Hong, M. (2002) Membrane protein topology probed by [1]h spin diffusion from lipids using solid-state NMR spectroscopy. *Journal of American Chemical Society*, **124**, 874-883.

37. Yamago, S., Iida, K., and Yoshida, J. (2002) Tailored synthesis of structurally defined polymers by organo tellurium-mediated living radical polymerization (TERP): synthesis of poly(meth)acrylate derivatives and their Di- and triblock copolymers. *Journal of American Chemical Society*, **124**, 13666-13667.

38. Yamago, S., Iida, K., Nakajima, M., Yoshida, J. (2003) Practical prot-

ocols for organo tellurium-mediated living radical polymerization by in situ generated initiators from AIBN and ditellurides. *Macromolecules*, **36**, 3793-3796.

39. Benedikt, S., Moszner, N., and Liska, R. (2014) Benzoyl phenyltelluride as highly reactive visible-light TERP reagent for controlled radical polymerization. *Macromolecules*, **47**, 5526-5531.

40. Mishima, E., and Yamago, S. (2011) Controlled alternating copolymerization of (meth)acrylates and vinyl ethers by using organoheteroatom-mediated living radical polymerization. *Macromolecular Rapid Communications*, **32**(12), 893-898.

41. Sugihara, Y., Kagawa, Y., Yamago, S., and Okubo, M. (2007) Organotellurium-mediated living radical polymerization in miniemulsion. *Macromolecules*, **40**, 9208-9211.

42. Okubo, M., Sugihara, Y., Kitayama, Y., Kagawa, Y., and Minami, H. (2009) Emulsifier-free, organotellurium-mediated living radical emulsion polymerization of butyl acrylate. *Macromolecules*, **42**, 1979-1984.

43. Inui, T., Yamanishi, K., Sato, E., and Matsumoto, A. (2013) Organotellurium-mediated living radical polymerization (TERP) of acrylates using ditelluride compounds and binary azo initiators for the synthesis of high-performance adhesive block copolymers for on-demand dismantlable adhesion. *Macromolecules*, **46**, 8111-8120.

44. Sugihara, Y., Yamago, S., and Zetterlund, P. B. (2015) An innovative approach to implementation of organotellurium- mediated radical polymerization (TERP) in emulsion polymerization. *Macromolecules*, **48**, 4312-4318.

45. Chiefari, J., Chong, Y. K., Ercole, F., Krstina, J., Jeffery, J., Le, T. P. T., Mayadunne, R. T. A., Meijs, G. F., Moad, C. L., Moad, G., Rizzardo, E., and Thang, S. H. (1998) Living free-radical polymerization by reversible addition-fragmentation chain transfer: The RAFT process. *Macromolecules*, **31**, 5559-5562.

46. Moad, G., Rizzardo, E., and Thang, S. H. (2009) Living radical polymerization by the RAFT process – A second ipdate. *Australian Journal of Chemistry*, **62**, 1402-1472.

47. McCormick, C. L., Lowe, A. B. (2004) Aqueous RAFT polymerization: recent developments in synthesis of functional water-soluble (Co) polymers with Controlled Structures. *Accounts of Chemical Research*, **37**, 312-325.

48. Mayadunne, R. T. A., Rizzardo, E., Chiefari, J., Chong, Y. K., Moad, G., Thang, S. H. (1999) Living radical polymerization with reversible addition–fragmentation chain transfer (RAFT polymerization) using dithiocarbamates as chain transfer agents. *Macromolecules*, **32**, 6977-6980.

49. Destarac, M., Charmot, D., Franck, X., and Zard, S. Z. (2000) Dithio-

carbamates as universal reversible addition-fragmentation chain transfer agents. *Macromolecular Rapid Communications*, **21**, 1035-1039.

50. Mayadunne, R. T. A., Rizzardo, E., Chiefari, J., Kristina, J., Moad, G., Postma, A., Thang, S. H. (2000) Living polymers by the use of tri-thiocarbonates as reversible addition–fragmentation chain transfer (RAFT) agents: ABA triblock copolymers by radical polymerization in two steps. *Macromolecules*, **33**, 243-245.

51. Francis, R., and Ajayaghosh, A. (2000) Minimization of homopolymer formation and control of dispersity in free radical induced graft polymerization using xanthate derived macro-photoinitiators. *Macromolecules*, **33**, 4699-4704.

52. Ladaviere, C., Dorr, N., and Claverie, J. P. (2001) Controlled radical polymerization of acrylic acid in protic media. *Macromolecules*, **34**, 5370-5372.

53. Ji, J., Jia, L., Yan, L., and Bangal, P. R. (2010) Efficient synthesis of poly (acrylic acid) in aqueous solution via a RAFT process. *Journal of Macromolecular Science, Part A: Pure and Applied Chemistry*, **47**, 445-451.

54. Barsbay, M., and Guven, O. (2009) A short review of radiation-induced raft-mediated graft copolymerization: a powerful combination for modifying the surface properties of polymers in a controlled manner. *Radiation Physics and Chemistry*, **78**, 1054-1059.

55. Hong, C. Y., You, Y. Z., Bai, R. K., Pan, C. Y., and Borjihan, G. (2001) Controlled polymerization of acrylic acid under 60Co irradiation in the presence of dibenzyl trithiocarbonate. *Journal of Polymer Science, Part A: Polymer Chemistry*, **39**, 3934-3939.

56. Millard, P. E., Barner, L., Stenzel, M. H., Davis, T. P., Kowollik, C. B., Muller, A. H. (2006) RAFT polymerization of N-isopropylacrylamide and acrylic acid under γ-irradiation in aqueous media. *Macromolecular Rapid Communications*, **27**, 821-828.

57. Millard, P. E., Barner, L., Reinhardt, J., Buchmeiser, M. R., Kowollik, C. B., and Muller, A. H. (2010) Synthesis of water-soluble homo-and block-copolymers by RAFT polymerization under γ-irradiation in aqueous media. *Polymer*, **51**, 4319-4328.

58. Klimkevicius, V., and Makuska, R. (2017) Successive RAFT polymerization of poly (ethylene oxide) methyl ether methacrylates with different length of PEO chains giving diblock brush copolymers. *European Polymer Journal*, **86**, 94-105.

59. Kulai, I., Brusylovets, O., Voitenko, Z., Harrisson, S., Mazieres, S., and Destarac, M. (2015) RAFT polymerization with triphenyl stannyl carbodithioates (Sn-RAFT). *ACS Macro Letters*, **4**, 809-813.

60. Chaduc, I., Lansalot, M., D'Agosto, F., and Charleux, B. (2012) RAFT polymerization of methacrylic acid in water. *Macromolecules*, **45**, 1241-1247.

61. Zhou, H., Liu, C., Qu, Y., Gao, C., Shi, K., and Zhang, W. (2016) How the polymerization procedures affect the morphology of the block copolymer nanoassemblies: Comparison between dispersion RAFT polymerization and seeded RAFT polymerization. *Macromolecules*, **49**, 8167-8176.

62. Nicolay, R. (2012) Synthesis of well-defined polythiol copolymers by RAFT polymerization. *Macromolecules*, **45**, 821-827.

63. Grover, G. N., Lee, J., Matsumoto, N. M., and Maynard, H. D. (2012) Aminooxy and pyridyl disulfide telechelic poly(poly(ethylene glycol) acrylate) by RAFT polymerization. *Macromolecules*, **45**, 4958-4965.

64. Minisci, F. (1975) Free-radical additions to olefins in the presence of redox systems. *Accounts of Chemical Research*, **8**, 165-171.

65. Wang, J. S., and Matyjaszewski, K. (1995) Free-radical additions to olefins in the presence of redox systems. *Journal of American Chemical Society*, **117**, 5614-5615.

66. Matyjaszewski, K. (1998) Inner sphere and outer sphere electron transfer reactions in atom transfer radical polymerization. *Macromolecular Symposia*, **134**, 105-118.

67. Singleton, D. A., Nowlan, D. T., Jahed, N., and Matyjaszewski, K. (2003) Isotope effects and the mechanism of atom transfer radical polymerization. *Macromolecules*, **36**, 8609-8616.

68. Fischer, H. (2001) The persistent radical effect: a principle for selective radical reactions and living radical polymerizations. *Chemical Reviews*, **101**, 3581-3610.

69. Tang, W., Tsarevsky, N. V., and Matyjaszewski, K. (2006) Determination of equilibrium constants for atom transfer radical polymerization. *Journal of the American Chemical Society*, **128**, 1598-1604.

70. Vana, P., Davis, T. P., and Kowollik, C. B. (2002) Easy access to chain-length-dependent termination rate coefficients using RAFT polymerization. *Macromolecular Rapid Communications*, **23**, 952-956.

71. Matyjaszewski, K., Coca, S., Gaynor, S. G., Wei, M., and Woodworth, B. E. (1998) Controlled radical polymerization in the presence of oxygen. *Macromolecules*, **31**, 5967-5969.

72. Braunecker, W. A., and Matyjaszewski, K. (2007) Controlled/living radical polymerization: Features, developments, and perspectives. *Progress in Polymer Science*, **32**, 93-146.

73. Tsarevsky, N. V., Braunecker, W. A., and Matyjaszewski, K. (2007) Electron transfer reactions relevant to atom transfer radical polymerization. *Journal of Organometallic Chemistry*, **692**, 3212-3222.

74. Shipp, D. A., Wang, J. L., and Matyjaszewski, K. (1998) Synthesis of acrylate and methacrylate block copolymers using atom transfer radical polymerization. *Macromolecules*, **31**, 8005-8008.

75. Robinson, K. L., Khan, M. A., Banez, M. V. P., Wang, X. S., and Armes, S. P. (2001) Controlled polymerization of 2-hydroxyethyl methacrylate by ATRP at ambient temperature. *Macromolecules*, **34**, 3155-3158.
76. Tong, Y. Y., Dong, Y. Q., Du, F. S., and Li, Z. C. (2008) Synthesis of well-defined poly(vinyl acetate)-b-polystyrene by combination of ATRP and RAFT polymerization. *Macromolecules*, **41**, 7339-7346.
77. Simakova, A., Averick, S. E., Konkolewicz, D., and Matyjaszewski, K. (2002) Aqueous ARGET ATRP. *Macromolecules*, **45**, 6371-6379.
78. Moad, G., and Solomon, D. H. (2006) *The Chemistry of Radical Polymerization*, 2nd edition, Elsevier, USA.
79. Moad, G., Chiefari, J., Chong, Y. K., Krstina, J., Mayadunne, R. T. A., Postma, A., Rizzardo, E., and Thang, S. H. (2000) Living free radical polymerization with reversible addition–fragmentation chain transfer (the life of RAFT). *Polymer International*, **49**, 993-1001.
80. Daikh, B. E., and Finke, R. G. (1992) The persistent radical effect: a prototype example of extreme, 105 to 1, product selectivity in a free-radical reaction involving persistent .cntdot.CoII[macrocycle] and alkyl free radicals. *Journal of the American Chemical Society*, **114**, 2938-2943.
81. Boyer, C., Bulmus, V., Davis, T. P., Ladmiral, V., Liu, J. Q., and Perrier, S. (2009) Bio applications of RAFT polymerization. *Chemical Reviews*, **109**, 5402-5436.
82. Monteiro, M. J. (2005) Modeling the molecular weight distribution of block copolymer formation in a reversible addition–fragmentation chain transfer mediated living radical polymerization. *Journal of Polymer Science, Part A: Polymer Chemistry*, **43**, 3189-3204.
83. Gossage, R. A., Kuil, L. A. V., and Koten, G. V. (1998) Diaminoarylnickel(II) "Pincer" complexes: mechanistic considerations in the kharasch addition reaction, controlled polymerization, and dendrimeric transition metal catalysts. *Accounts of Chemical Research*, **31**, 423-431.
84. Moad, G., Rizzardo, E., and Thang, S. H. (2006) Living radical polymerization by the RAFT process - A first update. *Australian Journal of Chemistry*, **59**, 669-692.
85. Moad, G., Ching, Y. K., Postma, A., Rizzardo, E., and Thang, S. H. (2005) Advances in RAFT polymerization: the synthesis of polymers with defined end-groups. *Polymer*, **46**, 8458-8468.
86. Derry, M. J., Fielding, L. A., and Armes, S. P. (2016) Polymerization-induced self-assembly of block copolymer nanoparticles via RAFT non-aqueous dispersion polymerization. *Progress in Polymer Science*, **52**, 1-18.
87. Ayres, N. (2011) Atom transfer radical polymerization: a robust and versatile route for polymer synthesis. *Polymer Reviews*, **51**,

138-162.

88. Ran, J., Wu, L., Zhang, Z., and Xu, T. (2014) Atom transfer radical polymerization (ATRP): A versatile and forceful tool for functional membranes. *Progress in Polymer Science*, **39**, 124-144.

89. Patten, T. E., and Matyjaszewski, K. (1998) Atom transfer radical polymerization and the synthesis of polymeric materials. *Advanced Materials*, **10**, 901-915.

90. Krys, P., and Matyjaszewski, K. (2017) Kinetics of atom transfer radical polymerization. *European Polymer Journal*, **89**, 482-523.

91. Ouchi, M., Terashima, T., and Sawamoto, M. (2008) Precision control of radical polymerization via transition metal catalysis: from dormant species to designed catalysts for precision functional polymers. *Accounts of Chemical Research*, **41** (9), 1120-1132.

92. Hawker, C. J., Bosman, A. W., and Harth, E. (2001) New polymer synthesis by nitroxide mediated living radical polymerizations. *Chemical Reviews*, **101**, 3661-3688.

93. Grubbs, R. B. (2011) Nitroxide-mediated radical polymerization: limitations and versatility. *Polymer Reviews*, **51**, 104-137.

94. Beejapur, H. A., Zhang, Q., Hu, K., Zhu, L., Wang, J., and Ye, Z. (2019) TEMPO in chemical transformations: From homogeneous to heterogeneous. *ACS Catalysis*, **9**(4), 2777-2830.

95. Chong, Y. K., Ercole, F., Moad, G., Rizzardo, E., Thang, S. H., and Anderson, A. G. (1999) Imidazolidinone nitroxide-mediated polymerization. *Macromolecules*, **32**, 6895-6903.

96. Chu, D. S. H., Schellinger, J. G., Shi, J., Convertine, A. J., Stayton, P. S., and Pun, S. H. (2012) Application of living free radical polymerization for nucleic acid delivery. *Macromolecules*, **45**, 1089-1099.

97. Li, B., Yu, B., Ye, Q., and Zhou, F. (2015) Tapping the potential of polymer brushes through synthesis. *Accounts of Chemical Research*, **48**, 229-237.

98. Zhou, F., Zheng, Z., Yu, B., Liu, W., Huck, W. T. S. (2006) Multicomponent polymer brushes. *Journal of American Chemical Society*, **128**, 16253-16258.

99. Baum, M., Brittain, W. J. (2002) Synthesis of polymer brushes on silicate substrates via reversible addition fragmentation chain transfer technique. *Macromolecules*, **35**, 610-615.

100. Ye, Q., Wang, X., Li, S., and Zhou, F. (2010) Surface-initiated ring-opening metathesis polymerization of pentadecafluorooctyl-5-norbornene-2- carboxylate from variable substrates modified with sticky biomimic initiator. *Macromolecules* **43**, 5554-5560.

101. Husseman, M., Malmstro, E. E., McNamara, M., Mate, M., Mecerreyes, D., Benoit, D. G., Hedrick, J. L., Mansky, P., Huang, E., Russell, T. P., and Hawker, C. J. (1999) Controlled synthesis of polymer brushes by "living" free radical polymerization techniques. *Macromolecules*, **32**, 1424-1431.

102. Wei, Q., Cai, M., Zhou, F., and Liu, W. (2013) Dramatically tuning friction using responsive polyelectrolyte brushes. *Macromolecules*, **46**, 9368-9379.

103. Wei, Q., Becherer, T., Uberti, S. A., Dzubiella, J., Wischke, C., Neffe, A. T., Lendlein, A., Ballauff, M., and Haag, R. (2014) Protein interactions with polymer coatings and biomaterials. *Angewandte Chemie International Edition*, **53**, 8004-8031.

104. Welch, M. E., Ritzert, N. L., Chen, H., Smith, N. L., Tague, M. E., Xu, Y., Baird, B. A., Abruna, H. D., and Ober, C. K. (2014) Generalized platform for antibody detection using the antibody catalyzed water oxidation pathway. *Journal of American Chemical Society*, **136**, 1879-1883.

105. Guo, R., Yu, Y., Xie, Z., Liu, X., Zhou, X., Gao, Y., Liu, Z., Zhou, F., Yang, Y., and Zheng, Z. (2013) Matrix-assisted catalytic printing for the fabrication of multiscale, flexible, foldable, and stretchable metal conductors. *Advanced Materials*, **25**, 3343-3350.

106. Jiang, H., and Xu, F. J. (2013) Biomolecule-functionalized polymer brushes. *Chemical Society Reviews*, **42**, 3394-3426.

107. Paripovic, D., and Klok, H. A. (2011) polymer brush guided formation of thin gold and palladium/gold bimetallic films. *ACS Applied Materials & Interfaces*, **3**, 910-917.

108. Wang, X., Guo, Q., Cai, X., Zhou, S., Kobe, B., and Yang, J. (2014) Initiator-integrated 3D printing enables the formation of complex metallic architectures. *ACS Applied Materials & Interfaces*, **6**, 2583-2587.

9

Polymer Nanocomposite Inks and Pigments

9.1 Introduction

Significant research attention has been drawn towards printed electronics in the recent years owing to usefulness in flexible, large size and cost-effective electronics [1-7]. Considerable progress has been made in printable semiconductor and electrode materials [7]. However, the dielectric layers, one of the most important constituents of thin-film transistors (TFTs), are not often printed due to the difficulty of printing a high-quality layer and the lack of printable dielectric materials [8]. Usually in TFTs, the thickness of dielectric films must be in the range of hundreds of nanometers to a few micrometers so as to avoid leakage current. Additionally, in order to bring down the gate voltage, it is required that the printed dielectric layer has large capacitance per unit area.

Various dielectric materials, based on solution processing, have been fabricated as gate dielectrics. These generally comprise of inorganic oxides [9], polymers [8], solid-state electrolytes [11] and ion-gels [10]. Solid state electrolytes and ion gels [6,7,12,13] have high dielectric constant and show enhanced conductivity at high frequency which restricts the transistor's switching speed. ZrO_2 [9] shows high frequency stability and permittivity, although the prevention of printed ceramic films from cracks and pinholes is difficult, leading to short circuits in appliances. However, polymer dielectric materials with low permittivity are relatively easier to print as compared to other materials. Composites integrating printable polymer and high-permittivity inorganic nanoparticles can result in novel dielectric materials having good printability, higher dielectric constant as well as high frequency stability. In such composite materials, the dielectric constant mostly elevates with the addition of high-permittivity inorganic nanoparticles. Nevertheless, with an increase in the concentration of nanoparticles, the leakage current enhances as well

Swati Singh and Vikas Mittal*, The Petroleum Institute (part of Khalifa University of Science and Technology), Abu Dhabi, UAE
*Current address: Bletchington, Wellington County, Australia

due to electromigration of the metallic atoms via gaps from the electrode material. In this case, two-dimensional (2D) dielectric materials are more suitable as compared to the spherical nanoparticles owing to their lamellar geometry [14-16]. Overall, various high performance polymers have been employed to develop nanocomposite inks and pigments for printing [17-38].

9.2 Printing Techniques

In three-dimensional printing, powdered materials are provided on a platform, and a liquid binder based on water is spread using inkjet printing so as to aggregate the individual particles and shape a two-dimensional sequence [39]. Another layer of powder is then introduced on the surface of the printed layers using a roller. In order to develop a strong bonding between the powder particles, post-heating is used [39]. A large variety of materials such as polymer composites, ceramics and metals can be printed using this method. Nevertheless, low printing resolution, high cost and non-optimal surface finish are concerns for three dimensional printing [40,41].

Laminated object printing is an amalgam of additive and subtractive manufacturing [42,43]. On the other hand, the mechanism of stereolithography is based on a photo-instigated polymerization [44-47]. Other printing techniques include paste extrusion printing, selective laser sintering, polyjet, laser engineered net shaping, etc. [48-70].

9.3 Polymer Nanocomposites based Inks

Peak *et al.* [71] investigated the rheological modification of poly(ethyleneglycol) (PEG) precursor solutions via incorporation of laponite clay nanoparticles. PEG/laponite formed internal "house-of-cards" structure, influencing fluid flow and ability to print (Figure 9.1). Laponite addition to PEG reduced the recovery time of solutions from infinite (Newtonian fluid) to seconds, which is more appropriate for bioprinting applications.

Titanium dioxide (TiO_2) nanoparticles are commonly employed in applications such as photocatalysts, batteries, electrochromic and photochromic devices, photovoltaic cells, gas sensors and thin-film transistors, due to the strong oxidizing power of the photogenerated holes, non-toxicity, high refractive index, chemical inertness and cost-effectiveness. N-doped TiO_2 nanoparticles were dissolved by

Loffredo *et al.* [72] in solution with the help of a polycation polymer (polyethyleneimine (PEI)) to prepare inks for ink jet printing applications. At different PEI concentrations, various TiO_2/PEI/EtOH suspensions were prepared, and the impact of the dispersant content on the ink printability was thoroughly investigated. The suspensions were observed to exhibit good chemico-physical behavior for their use as inks.

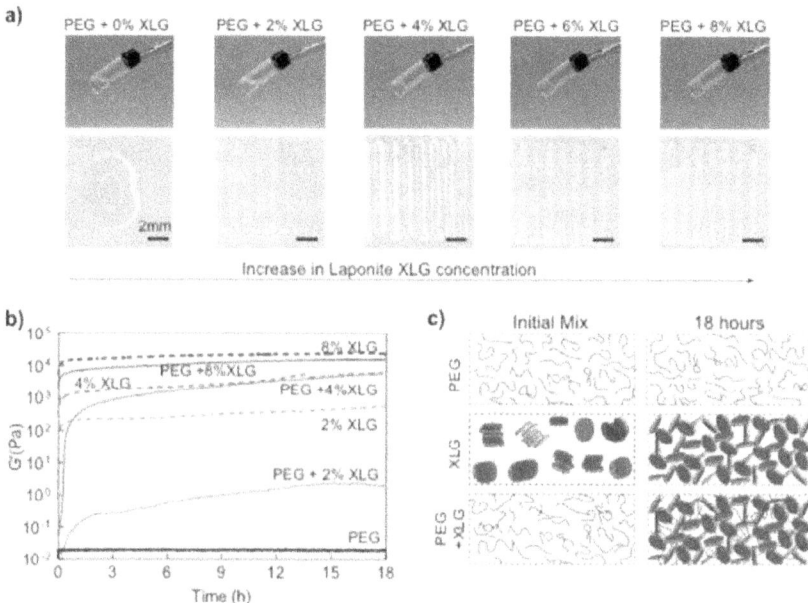

Figure 9.1 Synthesis of PEG-laponite colloidal solution: (a) viscosity increases at rest with increasing laponite concentration, (b) time sweep curves and (c) schematic of internal structure formation of PEG, laponite and PEG/laponite solutions. Reproduced from Reference 71 with permission from American Chemical Society.

In order to obtain an environment friendly and printable ink, a nanocomposite based on graphene and acrylic polymer was demonstrated [73]. Thin printing tests of the nanocomposite demonstrated a diminishing of resistivity by two orders of magnitude. Directly written TiO_2 films were fabricated by Arango *et al.* [74] on flexible substrates from aqueous systems. Mild temperature and UV irradiation conditions were used to transform the amorphous/crystalline formulations to semi-crystalline/crystalline films for diverse applications on flexible substrates. It was observed that the viscosity and printing properties of the inks could be tailored through solvent and polymer addition and titanium (IV) bis(ammonium lactato) dihydroxide (TALH):TiO_2 ratio (Figure 9.2).

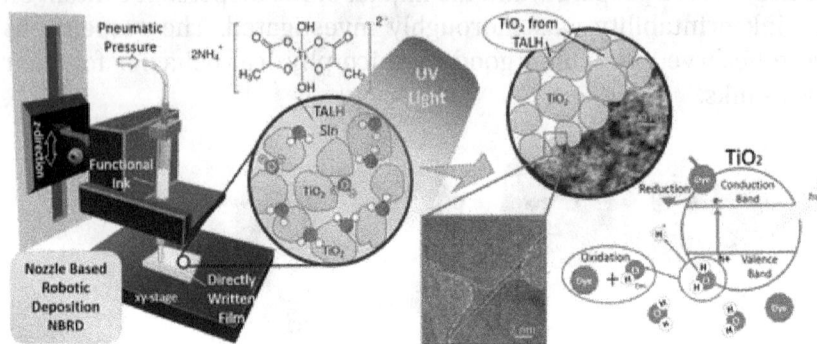

Figure 9.2 Schematic representation of the synthesis of printed films. Reproduced from Reference 74 with permission from American Chemical Society.

Natural composite materials were also developed over an expansive length scale by the immediate stage trans-development of all-fluid ink to composite material [75]. The variations in the degree of ink bead mixture amid printing were observed to create differing material morphologies.

9.4 Polymer Nanocomposites based Pigments

Natural shades have been widely utilized for paints and printing applications owing to their photosensitivity, shading quality, etc. Natural shades are hard to be wetted and dispersed in water or polymeric films [76,77]. In past decades, advancements in this area have been achieved through progression in heterogeneous polymerization and nanocomposites technology through emulsion polymerization [78,79], miniemulsion polymerization [80,81], suspension polymerization [82,83] or micro-suspension polymerization [84,85].

Qi *et al.* [86] reported improvement in dispersion and color of organic pigments in polymeric films via micro-encapsulation. Zeng *et al.* [87] also reviewed the applications of polymer nanocomposites in coatings and pigments. Films of PU/PVC containing multi-walled and single-walled carbon nanotubes (MWCNTs and SWCNTs) were prepared by Abdelrazek *et al.* [88]. In another study, nanocomposite films consisting of polyvinylidene fluoride (PVDF)/polyvinyl chloride (PVC) blend doped with 0.005 and 0.010 wt% of graphene oxide nanoparticles (GO) were reported [89].

Kiatkamjornwong and Pomsanam [83] reported styrenic-based polymerized toner and its composite for electrophotographic printing. The analysis of print quality exhibited high background fog, low maximum density and small extent of image raggedness. Nanoblue shade poly(styrene-co-n-butyl acrylate-co-methacrylic acid) composite particles were also reported by fast suspension polymerization by Widiyandari *et al.* [84]. In another study, simple and ecofriendly method was used to prepare titanium dioxide/polymer composites using compound oxidative polymerization [90]. Ghannam *et al.* [91] also reported colored diblock copolymer-mica composite pigments. The pigments were prepared from the mica surface by carrying out the block copolymerization of butyl acrylate and a mixture of styrene and a dye.

9.5 Conclusions

In this study, the development of polymer nanocomposite inks and pigments has been briefly reviewed. Owing to the much improved properties as compared to the conventional materials, these nanomaterials exhibit high potential of wide ranging applications in near future.

References

1. Wang, C., Qian, L., Xu, W., Nie, S., Gu, W., and Zhang, J. (2013) High performance thin film transistors based on regioregular poly(3-dodecylthiophene)-sorted large diameter semiconducting single-walled carbon nanotubes. *Nanoscale*, **5**, 4156-4161.
2. Han, S. Y., Lee, D. H., Herman, G. S., and Chang, C. H. (2009) Inkjet-printed high mobility transparent-oxide semiconductors. *Journal of Display Technology*, **5**, 520-524.
3. Yan, H., Chen, Z., Zheng, Y., Newman, C., Quinn, J. R., and Dötz, F. (2009) A high-mobility electron-transporting polymer for printed transistors. *Nature*, **457**, 679-686.
4. Woo, K., Bae, C., Jeong, Y., Kim, D., and Moon, J. (2010) Inkjet-printed Cu source/drain electrodes for solution-deposited thin film transistors. *Journal of Materials Chemistry*, **20**, 3877-3882.
5. Chung, S., Jang, M., Ji, S. B., Im, H., Seong, N., Ha, J., Kwon, S.-K., Kim, Y.-H., Yang, H., and Hong, Y. (2013) Flexible high-performance all-

inkjet-printed inverters: Organo-compatible and stable interface engineering. *Advanced Materials*, **25**, 4773-4777.

6. Zhao, J., Gao, Y., Gu, W., Wang, C., Lin, J., and Chen, Z. (2012) Fabrication and electrical properties of all-printed carbon nanotube thin film transistors on flexible substrates. *Journal of Materials Chemistry*, **22**, 20747-20753.

7. Zhao, J., Gao, Y., Lin, J., Chen, Z., and Cui, Z. (2012) Printed thin-film transistors with functionalized single-walled carbon nanotube inks. *Journal of Materials Chemistry*, **22**, 2051-2056.

8. Ortiz, R., Facchetti, A., and Marks, T. J. (2010) High-k organic, inorganic, and hybrid dielectrics for low-voltage organic field-effect transistors. *Chemical Reviews*, **110**, 205-239.

9. Park, J. H., Yoo, Y. B., Lee, K. H., Jang, W., Oh, J., and Chae, S. (2012) Low-temperature, high- performance solution-processed thin-film transistors with peroxo-zirconium oxide dielectric. *ACS Applied Materials & Interfaces*, **5**, 410-417.

10. Cho, J. H., Lee, J., He, Y., Kim, B. S., Lodge, T. P., and Frisbie, C. D. (2008) High-capacitance ion gel gate dielectrics with faster polarization response times for organic thin film transistors. *Advanced Materials*, **20**, 686-690.

11. Braga, D., Ha, M., Xie, W., and Frisbie, C. D. (2010) Ultralow contact resistance in electrolytegated organic thin film transistors. *Applied Physics Letters*, **97**, 193311.

12. Lee, J., Kaake, L. G., Cho, J. H., Zhu, X. Y., Lodge, T. P., and Frisbie, C. D. (2009) Ion gel-gated polymer thin-film transistors: operating mechanism and characterization of gate dielectric capacitance, switching speed, and stability. *Journal of Physical Chemistry C*, **113**, 8972-8981.

13. Hong, K., Kim, S. H., Lee, K. H., and Frisbie, C. D. (2013) Printed sub-2V ZnO electrolyte gated transistors and inverters on plastic. *Advanced Materials*, **25**, 3413-3418.

14. Jang, W. S., Rawson, I., and Grunlan, J. C. (2008) Layer-by-layer assembly of thin film oxygen barrier. *Thin Solid Films*, **516**, 4819-4825.

15. Wu, X., Fei, F., Chen, Z., Su, W., and Cui, Z. (2014) A new nanocomposite dielectric ink and its application in printed thin-film transistors. *Composites Science and Technology*, **94**, 117-122.

16. *Polymer Nanocomposites, Printable and Flexible Technology for Electronic Packaging*. Online: https://smtnet.com/library/files/upload/Printable-Nanocomposites.pdf [accessed 26th June 2019].

17. Yang, X., Loos, J., Veenstra, S. C., Verhees, W. J. H., Wienk, M. M., Kroon, J. M., Michels, M. A. J., and Janssen, R. A. J. (2005) Nanoscale morphology of high-performance polymer solar cells. *Nano Letters*, **5**, 580-582.

18. *High Performance Polymers and Engineering Plastics*, Mittal, V. (ed.),

John Wiley & Sons, USA (2011).

19. Hiemenz P. C., and Lodge, T. P. (2007) *Polymer Chemistry*, CRC Press, USA.

20. Cogswell, F. N. (2013) *Thermoplastic Aromatic Polymer Composites*, Elsevier, USA.

21. Molazemhosseini, A., Tourani, H., Naimi-Jamal, M. R., and Khavandi, A. (2013) Nanoindentation and nanoscratching responses of PEEK based hybrid composites reinforced with short carbon fibers and nano-silica. *Polymer Testing*, **32**, 525-534.

22. Díez-Pascual, A. M., Naffakh, M., Gonzalez-Domínguez, J. M., Anson, A., Martínez-Rubi, Y., Martinez, M. T., Simard, B., Gomez, M. A. (2010) High performance PEEK/carbon nanotube composites compatibilized with polysulfones-II. Mechanical and electrical properties. *Carbon*, **48**, 3500-3511.

23. LiuJie, X., Davim, J. P., Cardoso, R. (2007) Prediction on tribological behaviour of composite PEEK-CF30 using artificial neural networks. *Journal of Materials Processing Technology*, **189**, 374-378.

24. Nisa, V., Rajesh, S., Murali, K., Priyadarsini, V., Potty, S., and Ratheesh, R. (2008) Preparation, characterization and dielectric properties of temperature stable $SrTiO_3$/PEEK composites for microwave substrate applications. *Composites Science and Technology*, **68**, 106-112.

25. Iqbal, T., Briscoe, B., and Luckham, P. (2011) Scratch deformations of poly (etheretherketone). *Wear*, **271**, 1181-1193.

26. Iqbal, T., Briscoe, B. J., Yasin, S., and Luckham, P. F. (2013) Nanoindentation response of poly(ether ether ketone) surfaces da semicrystalline bimodal behavior. *Journal of Applied Polymer Science*, **130**, 4401-4409.

27. Powles, R., McKenzie, D., Meure, S., Swain, M., and James, N. (2007) Nanoindentation response of PEEK modified by mesh-assisted plasma immersion ion implantation. *Surface and Coatings Technology*, **201**, 7961-7969.

28. Díez-Pascual, A. M., Gomez-Fatou, M. A., Ania, F., and Flores, A. (2015) Nanoindentation in polymer nanocomposites. *Progress in Materials Science*, **67**, 1-94.

29. Godara, A., Raabe, D., and Green, S. (2007) The influence of sterilization processes on the micromechanical properties of carbon fiber-reinforced PEEK composites for bone implant applications. *Acta Biomaterialia*, **3**, 209-220.

30. Voyiadjis, G. Z., Samadi-Dooki, A., Malekmotiei, L. (2017) Nanoindentation of high performance semicrystalline polymers: A case study on PEEK. *Polymer Testing*, **61**, 57-64.

31. *Functional Polymer Blends*, Mittal, V. (ed.), CRC Press, USA (2012).

32. *Advances in Polyolefins*, Seymour, R. B., and Cheng, T. C. (eds.), Springer, USA (1987).

33. Schnell, H. (1956) Polycarbonate, eine gruppe neuartiger thermo-plastischer kunststoffe. herstellung und eigenschaften aromatischer polyester der kohlensäure. *Angewandte Chemie*, **68**(2), 633-640.

34. *Applied Polymer Science*, Craver, J. K., and Tess, R. W. (eds.), American Chemical Society, USA (1975).

35. De Leon, A. C., Chen, Q., Palaganas, N. B., Palaganas, J. O., Manapat, J., and Advincula, R. C. (2016) High performance polymer nanocomposites for additive manufacturing applications. *Reactive and Functional Polymers*, **103**, 141-155.

36. Carothers W. H. (1938) Diamine-dibasic Acid Salts, patent US2130947.

37. *Polymer Brushes*, Mittal, V. (ed.), CRC Press, USA (2012).

38. *Miniemulsion Polymerization Technology*, Mittal, V. (ed.), John Wiley, USA (2010).

39. Ligon, S. C., Liski, R., Stampfl, J., Gurr, M., and Mülhaupt, R. (2017) Polymers for 3D printing and customized additive manufacturing. *Chemical Reviews*, **117**, 10212-10290.

40. Lee, K. S., Kim, R. H., Yang, D. Y., and Park, S. H. (2008) Advances in 3D nano/microfabrication using two-photon initiated polymerization. *Progress in Polymer Science*, **33**, 631-681.

41. *Old World Labs*. Online: https://www.oldworldlabs.com [accessed 12th June 2019].

42. Mudge, R. P., and Wald, N.R. (2007) Laser engineered net shaping advances additive manufacturing and repair. *Welding Journal*, **86**, 44-48.

43. Singh, R. (2011) Process capability study of polyjet printing for plastic components. *Journal of Mechanical Science and Technology*, **25**, 1011-1015.

44. *Fast, Precise, Safe Prototype with FDM*. Online: http://sffsymposium.engr.utexas.edu/Manuscripts/1991/1991-15-Wales.pdf [accessed 19th June 2019].

45. Noorani, R. (2006) *Rapid Prototyping: Principles and Applications*, John Wiley & Sons, USA.

46. Gibson, I., Rosen, D., and Stucker, B. (2015) *Additive Manufacturing Technologies*, 2nd edition, Springer, USA.

47. Sciancaleporea, C., Moroni, F., Messoria, M., and Bondioli, F. (2017) Acrylate-based silver nanocomposite by simultaneous polymerization–reduction approach via 3D stereolithography. *Composites Communications*, **6**, 11-16.

48. Mota, C., Puppi, D., Dinucci, D., Gazzarri, M., and Chiellini, F. (2013) Additive manufacturing of star poly(ε-caprolactone) wet-spun scaffolds for bone tissue engineering applications. *Journal of Bioactive and Compatible Polymers*, **28**, 320-340.

49. Muth, J. T., Vogt, D. M., Truby, R. L., Mengüç, Y., Kolesky, D. B., Wood,

R. J., and Lewis, J. A. (2014) Embedded 3D printing of strain sensors within highly stretchable elastomers. *Advanced Materials*, **26**, 6307-6312.

50. Le, H. P. (1998) Progress and trends in ink-jet print technology. *Journal of Imaging Science and Technology*, **42**, 49-62.

51. Feygin, M., and Hsieh, B. (1991) Laminated Object Manufacturing (LOM): A Simpler Process. *1991 International Solid Freeform Fabrication Symposium*.

52. Kamrani, A. K., and Nasr, E. A. (2010) *Engineering Design and Rapid Prototyping*, Springer, USA.

53. Sachs, E., Cima, M., and Cornie, J. (1990) Three dimensional printing: rapid tooling and prototypes directly from a CAD model. *CIRP Annals*, **39**, 201-204.

54. Huang, S. H., Liu, P., Mokasdar, A., and Hou, L. (2013) Additive manufacturing and its societal impact: A literature review. *The International Journal of Advanced Manufacturing Technology*, **67**, 1191-1203.

55. Kim, G. D., and Oh, Y. T. (2008) A benchmark study on rapid prototyping processes and machines: quantitative comparisons of mechanical properties, accuracy, roughness, speed, and material cost. *Proceedings of the Institution of Mechanical Engineers, Part B*, **22**, 201-215.

56. Deckard, C., and Beaman, J. J. (1988) Process and control issues in selective laser sintering. *Sensors and Controls for Manufacturing*, USA, pp. 191-197.

57. Beaman, J. J., Barlow, J. W., Bourell, D. L., Crawford, R. H., Marcus, H. L., and McAlea, K. P. (1996) *Solid Freeform Fabrication: A New Direction in Manufacturing*, Springer, USA.

58. Murr, L. E., Gaytan, S. M., Ramirez, D. A., Martinez, E., Hernandez, J., Amato, K. N., Shindo, P. W., Medina, F. R., Wicker, R. B. (2012) Metal fabrication by additive manufacturing using laser and electron beam melting technologies. *Journal of Material Science and Technology*, **28**, 1-14.

59. Tang, H. H., Chiu, M. L., and Yen, H. C. (2011) Slurry based selective laser sintering of polymer-coated ceramic powders to fabricate high strength alumina parts. *Journal of European Ceramic Society*, **31**, 1383-1388.

60. Salentijn, G. I. J., Oome, P. E., Grajewski, M., and Verpoorte, E. (2017) Fused deposition modeling 3d printing for (bio)analytical device fabrication: procedures, materials, and applications. *Analytical Chemistry*, **89**, 7053-7061.

61. Pham, D. T., Ji, C. (2000) Design for stereolithography. *Proceedings of the Institution of Mechanical Engineers*, **214**, 635-640.

62. Melchels, F. P. W., Feijen, J., Grijpma, D. W. (2010) A review on stereolithography and its applications in biomedical engineering. *Bio-*

materials, **31**, 6121-6130.

63. Therriault, D., Shepherd, R. F., White, S. R., and Lewis, J. A. (2005) Fugitive inks for direct-write assembly of three-dimensional micro-vascular networks. *Advanced Materials*, **17**, 395-399.

64. Duoss, E. B., Weisgraber, T. H., Hearon, K., Zhu, C., Small, W., Metz, T. R., and Spadaccini, C. M. (2014) Three-dimensional printing of elastomeric, cellular architectures with negative stiffness. *Advanced Functional Materials*, **24**, 4905-4913.

65. Nelson, A., and Cosgrove, T. (2004) Dynamic light scattering studies of poly(ethylene oxide) adsorbed on laponite: Layer conformation and its effect on particle stability. *Langmuir*, **20**, 10382-10388.

66. Schmidt, G., Nakatani, A. I., and Han, C. C. (2002) Rheology and flow-birefringence from viscoelastic polymer-clay solutions. *Rheologica Acta*, **41**, 45-54.

67. Nelson, A., and Cosgrove, T. (2004) A small-angle neutron scattering study of adsorbed poly(ethylene oxide) on laponite. *Langmuir*, **20**, 2298-2304.

68. Mohanty, R. P., Suman, K., and Joshi, Y. M. (2017) In situ ion induced gelation of colloidal dispersion of Laponite: Relating microscopic interactions to macroscopic behavior. *Applied Clay Science*, **138**, 17-24.

69. Zulian, L., Ruzicka, B., and Ruocco, G. (2008) Influence of an adsorbing polymer on the aging dynamics of Laponite clay suspensions. *Philosophical Magazine*, **88**, 4213-4221.

70. Zulian, L., Marques, F. A. D., Emilitri, E., Ruocco, G., and Ruzicka, B. (2014) Dual aging behavior in a clay-polymer dispersion. *Soft Matter*, **10**, 4513-4521.

71. Peak, C. W., Stein, J., Gold, K. A., and Gaharwar, A. K. (2017) Nanoengineered colloidal inks for 3D bioprinting. *Langmuir*, **34**(3), 917-925.

72. Loffredo, F., Grimaldi, I. A., Mauro, A. D. G. D., Villani, F., Bizzarro, V., Nenna, G., D'Amato, R., and Minarini, C. (2011) Polyethylenimine/N-Doped titanium dioxide nanoparticlebased inks for ink-jet printing applications. *Journal of Applied Polymer Science*, **122**, 3630-3636.

73. Giardi, R., Porro, S., Chiolerio, A., Celasco, E., and Sangermano, M. (2013) Inkjet printed acrylic formulations based on UV-reduced graphene oxide nanocomposites. *Journal of Materials Science*, **48**, 1249-1255.

74. Arango, M. A. T., Andrade, A. S. V., Cipollone, D. T., Grant, L. O., Korakakis, D., and Sierros, K. A. (2016) Robotic deposition of TiO_2 films on flexible substrates from hybrid inks: Investigation of synthesis-processing-microstructure-photocatalytic relationships. *ACS Applied Materials & Interfaces*, **8**, 24659-24670.

75. Hayashi, K., Morii, H., Iwasaki, K., Horie, S., Horiishi, N., and Ichimura, K. (2007) Uniformed nano-downsizing of organic pigme-

nts through core-shell structuring. *Journal of Materials Chemistry*, **17**, 527-530.

76. Fu, S., Ding, L., Xu, C., and Wang, C. (2010) Properties of copper phthalocyanine blue encapsulated with a copolymer of styrene and maleic acid. *Journal of Applied Polymer Science*, **117**, 211-215.

77. Fu, S., Xu, C., Du, C., Tian, A., and Zhang, M. (2011) Encapsulation of C.I. Pigment blue 15:3 using a polymerizable dispersant via emulsion polymerization. *Colloids and Surfaces A*, **384**, 68-74.

78. Nguyen, D., Zondanos, H. S., Farrugia, J. M., Serelis, A. K., Such, C. H., and Hawkett, B. S. (2008) Pigment encapsulation by emulsion polymerization using macro-RAFT copolymers. *Langmuir*, **24**, 2140-2150.

79. Lelu, S., Novat, C., Graillat, C., Guyot, A., and Bourgeat-Lami, E. (2003) Encapsulation of an organic phthalocyanine blue pigment into polystyrene latex particles using a miniemulsion polymerization process. *Polymer International*, **52**, 542-547.

80. Steiert, N., and Landfester, K. (2007) Encapsulation of organic pigment particles via miniemulsion polymerization. *Macromolecular Materials and Engineering*, **292**, 1111-1125.

81. Fu, S. H., and Fang, K. J. (2007) Preparation of styrene-maleic acid copolymers and its application in encapsulated pigment red 122 dispersion. *Journal of Applied Polymer Science*, **105**, 317-321.

82. Yang, J., Wang, T. J., He, H., Wei, F., and Jin, Y. (2003) Particle size distribution and morphology of in situ suspension polymerized toner. *Industrial and Engineering Chemistry Research*, **42**, 5568-5575.

83. Kiatkamjornwong, S., and Pomsanam, P. (2003) Synthesis and characterization of styrenic-based polymerized toner and its composite for electrophotographic printing. *Journal of Applied Polymer Science*, **89**, 238-248.

84. Widiyandari, H., Iskandar, F., Hagura, N., and Okuyama, K. (2008) Preparation and characterization of nanopigment-poly(styrene-co-w-butyl acrylate-co-methacrylic acid) composite particles by high speed homogenization-assisted suspension polymerization. *Journal of Applied Polymer Science*, **108**, 1288-1297.

85. Qi, D., Zhang, R., Xu, L., Yuan, Y., and Lei, L. (2011) Preparation and characterization of organic pigment phthalocyanine blue microcapsules by in-situ micro-suspension polymerization. *Acta Polymerica Sinica*, **11**, 145-150.

86. Qi, D., Chen, Z., Yang, L., Cao, Z., and Wu, M. (2013) Improvement of dispersion and color effect of organic pigments in polymeric films via microencapsulation by the miniemulsion technique. *Advances in Materials Science and Engineering*, **2013**, doi: 10.1155/2013/790321.

87. Zeng, Q. H., Yu, A. B., Lu, G. Q., and Paul, D. R. (2005) Clay-based pol-

ymer nanocomposites: research and commercial development. *Journal of Nanoscience and Nanotechnology*, **5**, 1574-1592.

88. Abdelrazek, E. M., Elashmawi, I. S., Hezma, A. M., Rajeh, A., and Kamal, M. (2016) Effect of an encapsulate carbon nanotubes (CNTs) on structural and electrical properties of PU/PVC nanocomposites. *Physica B*, **502**, 48-55.

89. Elashmawi, I. S., Alatawi, N. S., and Elsayed, N. H. (2017) Preparation and characterization of polymer nanocomposites based on PVDF/PVC doped with graphene nanoparticles. *Results in Physics*, **7**, 636-640.

90. Jadhav, N., and Gelling, V. (2014) Titanium dioxide/conducting polymers composite pigments for corrosion protection of cold rolled steel. *Journal of Coatings Technology and Research*, **12**, 137-152.

91. Ghannam, L., Garay, H., Shanahan, M. E. R., Francüois, J., and Billon, L. (2005) A new pigment type: Colored diblock copolymer-mica composites. *Chemistry of Materials*, **17**, 3837-3843.

Index

C

■ Index

P

packaging, 1, 3, 95, 105, 114, 130

PANI, 94

particle size, 135

percolation, 16-17, 32-33

permeability, 55-57, 88

permeation, 16, 33, 55-56, 70

phase behavior, 10

photocatalyst, 126

photocatalytic, 80, 96, 134

photoirradiation, 117

PLA, 95, 132

plasticization, 71

PMMA, 35, 112

polydispersity, 109-110

polymerization, 3, 17, 34-35, 37, 47, 52, 54, 74, 81-82, 87, 91, 93, 99, 104, 107-122, 126, 128-129, 132, 135

polypropylene, 3, 16-17, 32-34, 104

polystyrene, 17, 34, 97, 113, 117, 121, 135

polyurethane, 37, 45, 98

pore size, 38, 56

pore volume, 43

porosity, 37-38, 45, 47, 52

porous network, 55

processability, 1, 55, 73

PVC, 93, 128-129, 136

R

random copolymer, 97-98, 103

re-crystallization, 5-6, 22, 25

recovery, 126

reduced graphene oxide, 83, 134

reflection, 5, 8, 22, 27, 49, 58-59, 75

refractive index, 5, 22, 126

regeneration, 38-41

relaxation, 16, 33

response time, 130

reverse osmosis, 73

rheology, 17, 34, 134

ring-opening metathesis, 114, 122

S

scattering, 46, 78, 134

W